Plastic Theory of Structures

Plastic Theory of Structures

Michael R Horne

The M.I.T. Press
Cambridge, Massachusetts

Copyright © Michael R. Horne 1971
First published in Great Britain 1971
by Thomas Nelson and Sons Ltd

First M.I.T. Press Edition 1971

ISBN 0 262 08050 8 (hardcover)

Library of Congress catalog card number: 73–175017

Made and printed in Great Britain by
William Clowes & Sons, Limited, London, Beccles and Colchester

Contents

Preface

The plastic theory of structures is an essential complement to elastic theory. The readily applied forms of either theory depend on idealized mathematical models both of the material properties and of the fabricated nature of the structure. By applying each theory in turn to the same structure, a much better overall picture is obtained of the stability, rigidity, and strength than is possible by applying either theory on its own.

An important feature of plastic theory is the extent to which intuitive ideas of structural behaviour may be used to help in the derivation of solutions. It has been the aim in this volume to make full use of such intuitive ideas in conveying an understanding of the principles involved. The intuitive approach is followed by formal statements and proofs of the theorems of plasticity, after which there is an exposition of the methods of plastic analysis. Emphasis is placed on methods of analysis which are of value in design applications, since it is in relation to design that plastic theory is of the greatest interest. The methods described are those suited primarily to hand calculations, since a thorough understanding of plastic theory is best attained by the working of examples, but some methods suitable for computer application are also included. Considerable care has been taken in providing suitable problems at the end of each chapter and answers are given at the end of the volume.

In addition to three chapters on the theory of plastic collapse, a chapter is devoted to minimum weight design and another to shakedown analysis. In these chapters, the same general pattern of exposition is adopted—that is, first to introduce the subject by specific examples, then to quote and prove the relevant theorems, and finally to present systematic methods of solution. The final chapter deals with the problem of stability. This is a subject in which, for an adequate understanding of structural behaviour, elastic and plastic theories of ultimate strength must be taken together. Only an introduction to this quite difficult subject is possible, but some treatment is essential if a balanced view of the significance of stability theory in relation to plastic theory is to be achieved.

An understanding of the formal proofs of the theorems of plastic collapse, minimum weight design, and shakedown is not essential to an overall understanding of plastic theory and these proofs may therefore be omitted on a first reading. A bibliography is given at the end of each chapter.

The author would like to thank Mr. L. J. Morris for some material on joints, and Professor E. R. Bryan for reading and commenting on the manuscript.

M. R. H.

Chapter 1
Plastic failure

1.1 Introduction

This volume is concerned with the behaviour beyond the elastic limit (with particular emphasis on the failure loads) of structures in which resistance to bending action is the primary means by which the loads are supported. Examples of such structures (simply supported and continuous beams and rigid frames) are shown in Figs 1.1(a), (b), and (c). In triangulated structures loaded at the nodes only [Fig. 1.1(d)], the loads are supported primarily by axial forces, and such structures will not be considered. In beams, shear forces are also present, while in rigid frames both shear and axial forces exist in at least some of the members. Except in tall frames, the greater stresses in the elastic range are those due to bending, and it is therefore not surprising that the ultimate resistance to bending is the most important factor in determining failure loads. For this reason the effect of shear and axial forces will in the first instance be neglected. Their effect is discussed in Chapter 3.

It is usual when developing elastic theories for structures to neglect in the first instance the effect of the change of geometry on the equations of equilibrium (Horne[1]). The resulting linear theory is found to be adequate as a first approximation except for slender members and structures, for which elastic stability becomes an important consideration (Horne and Merchant,[2] Smith and Sidebottom[3]). The increase of deflections beyond the elastic limit may mean that the neglect of change of geometry is then less justifiable, but it is still found that useful and significant calculations can be made by adopting this same approximation. The effects of change of geometry and of instability are discussed briefly in Chapter 6, although a detailed treatment is beyond the scope of this volume.

The behaviour of materials beyond the elastic limit is highly variable, and it is intuitively obvious that no theory which attempted to deal accurately with structures composed of continuously strain-hardening materials [Fig. 1.2(a)] would lead to easily manageable results. The simple *elastic–pure plastic* stress–strain relation OAB shown in Fig. 1.2(b) leads to easier calculations and represents a close approximation to the behaviour of mild steel as well as a useful

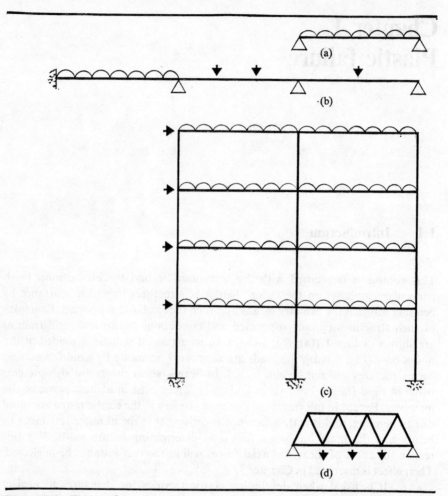

Figure 1.1   Types of structure

first approximation to some continuously strain-hardening materials. Pure plastic deformation up to indefinitely high strains is assumed to be possible at the yield stress σ_y, but a reversal in the direction of straining [CD in Fig. 1.2(b)] is accompanied by unloading with change of stress related to change of strain as in the elastic range. The stress–strain relation is assumed to be identical in tension and compression, and reversal of stress after plastic deformation has occurred is assumed to remain elastic until the yield stress is reached in the opposite sense [CDA′ in Fig. 1.2(b)]. This latter assumption ignores the Bauschinger effect observed in practice.

Although the elastic–pure plastic stress–strain relation is a reasonable approximation to the behaviour of some materials, it is strictly speaking no more than a mathematical model, ultimately to be judged by the extent to which it leads to useful results. A mathematical model having many theoretical advantages is the *rigid–plastic* model [Fig. 1.2(c)], equivalent to Fig. 1.2(b) with

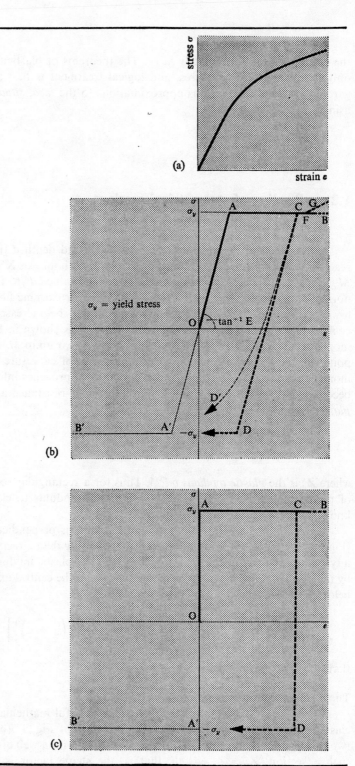

Figure 1.2   Stress–strain relationships

the elastic modulus E infinitely large. The theorems of plasticity apply strictly only to rigid–plastic structures, and logical treatment is best achieved by regarding *practical situations* as approximations to the *ideal situations* solved by using the theorems.

1.2   Bending of symmetrical sections

A beam of rectangular section, width b and depth d [Fig. 1.3(a)], will reach its elastic limit under a sagging *yield moment* (about axis XX) of magnitude $M_y = Z_e \sigma_y$, where Z_e, the *elastic modulus*, has the value $bd^2/6$. The longitudinal normal stress varies linearly from $-\sigma_y$ in the upper extreme fibres to σ_y in the lower extreme fibres. The bending moment may be increased beyond M_y, causing plastic zones in tension and compression as shown in Fig. 1.3(b). The moment–curvature relation (OAB in Fig. 1.4) is linear up to M_y (represented by point A), but shows an increasing rate of change of curvature with change of moment as plasticity develops. As the curvature approaches infinity the section becomes fully plastic [Fig. 1.3(c)], the moment of resistance approaching the *plastic moment*

$$M_p = \left\{ \left(\frac{bd}{2} \sigma_y \right) \frac{d}{4} \right\} \times 2 = Z_p \sigma_y$$

where Z_p is the *plastic modulus* $bd^2/4$. Thus for a rectangular section, the ratio of plastic moment to yield moment (or plastic modulus to elastic modulus), known as the shape factor and denoted by ν, is 1·5.

A symmetrical I-section, bent about an axis perpendicular to the web [Fig. 1.5(a)], has a shape factor much nearer to unity than a rectangular section, a typical value for a rolled section being 1·15. The plastic modulus is calculated by taking the moment of area of the section about the central axis XX, the area below XX being regarded as negative. Hence

$$Z_p = 2 \left[\{bt_f\}\left\{ \frac{d}{2} - \frac{t_f}{2} \right\} + \left\{ t_w \left(\frac{d}{2} - t_f \right) \right\}\left\{ \frac{1}{2} \left(\frac{d}{2} - t_f \right) \right\} \right]$$

that is, $Z_p = bt_f (d - t_f) + \frac{1}{4} t_w (d - 2t_f)^2$ (1.1)

The moment–curvature relation is OA′B′ in Fig. 1.4

An idealized section which provides a useful mathematical model for elastic–plastic behaviour is the twin-plate section of Fig. 1.5(b), consisting of two thin plates each of area $A/2$, distance d apart, with a web of zero area. For such a section, $Z_e = Z_p = \frac{1}{2}Ad$, that is, the shape factor is unity, and the moment–curvature relationship becomes OA″B″ in Fig. 1.4.

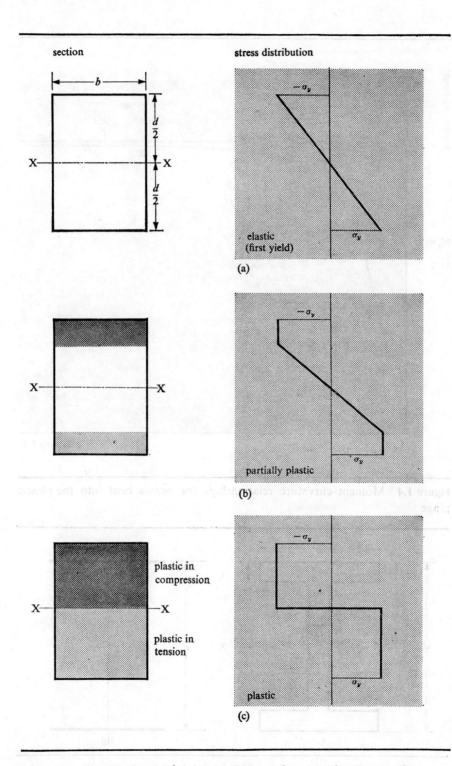

Figure 1.3 Bending into plastic range of beam of rectangular cross section

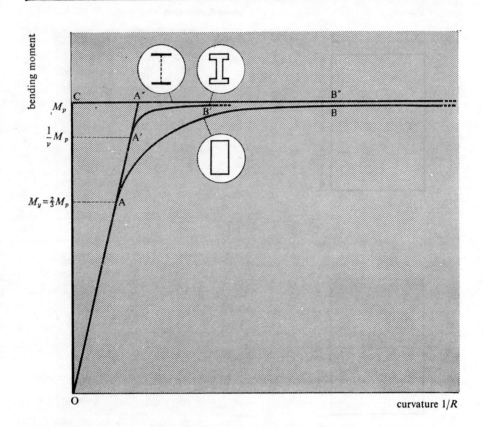

Figure 1.4   Moment–curvature relationships for beams bent into the plastic range

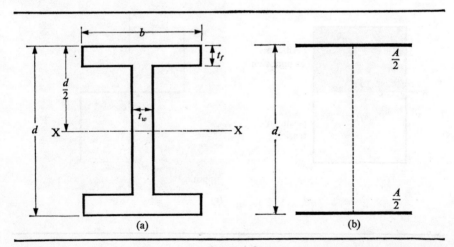

Figure 1.5   (a) Beam of I-section; (b) 'twin-plate' or unit shape factor section

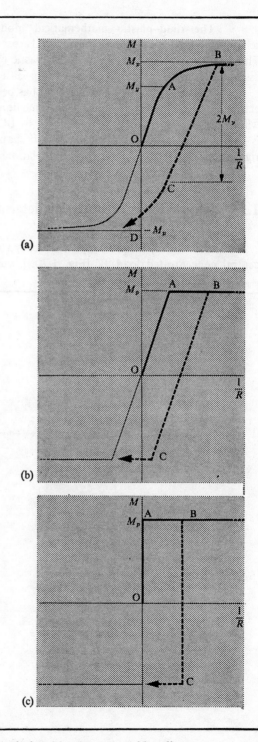

Figure 1.6 Moment–curvature relationships for reversed bending

The rigid–plastic mathematical model for the moment–curvature relationship is the same for members of all cross sections, being OCB″ in Fig. 1.4. The member remains straight until the plastic moment is reached, and can then assume any curvature in the corresponding sense.

The unloading of a beam after it has become partially plastic follows an elastic relationship over a change of extreme fibre stress of $2\sigma_y$. Hence, unloading will be linear over a range $2M_y$ as shown by BC in Fig. 1.6(a). The linear range is $2M_p$ for an elastic–plastic beam of unit shape factor [Fig. 1.6(b)], and for a rigid–plastic beam the moment reversal relationship is as in Fig. 1.6(c).

1.3   Failure loads of statically determinate structures

A uniform, simply supported beam with full plastic moment M_p and central concentrated load W [Fig. 1.7(a)] will have a linear load–deflection

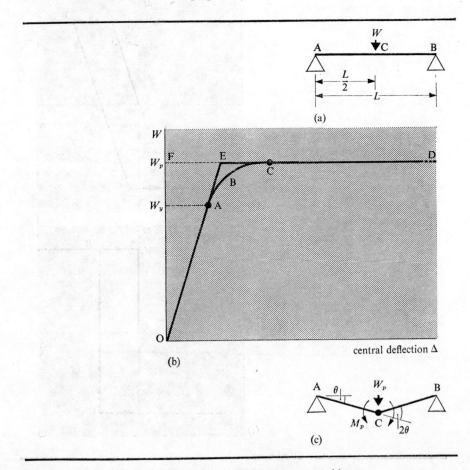

Figure 1.7   Elastic–plastic behaviour of simply supported beam

relation OA [Fig. 1.7(b)] until the central moment becomes M_y. The corresponding load W_y is statically determined by taking moments about C for AC, giving $M_y = W_y L/4$ or $W_y = 4M_y/L$. With increase of load, plastic zones will spread from the centre of length of the beam until full plasticity occurs under the load at a value $W_p = 4M_p/L$ [point C in Fig. 1.7(b)]. Thereafter, assuming infinite strain capacity at the yield stress in tension and compression, infinitely high curvature occurs at the load and the deflection can increase indefinitely [CD in Fig. 1.7(b)]. We therefore postulate the formation of a *plastic hinge* of constant moment of resistance M_p under the load. Since the moments in the beam are everywhere constant as deformation proceeds, *increments* of deformation are correctly represented by the rigid-link mechanism in Fig. 1.7(c). Equating the external work to the internal work absorbed as plastic deformation in the hinge, and assuming θ to be small,

$$W_p \left(\frac{L\theta}{2} \right) = M_p(2\theta) \tag{1.2}$$

giving $W_p = 4M_p/L$ as before. Equation (1.2) may be regarded either as a *real* or a *virtual* work equation.

The concept of a plastic hinge is obviously a physical approximation, since infinite curvature is impossible without complete fracture of the beam. More correctly, high curvature occurs over a short length on either side of the section of maximum moment, this being made possible by the increase of moment of resistance at this section to a value in excess of the plastic moment, thus enabling plasticity to spread (Horne,[4] Hrennikoff[5]). It is found experimentally, however, that for mild steel structures the plastic hinge concept represents a close approximation to the truth.

If unit shape factor is assumed, the load–deflection relation becomes the two straight lines OED. A rigid–plastic beam has the relationship OFD, and in this case the rigid-link mechanism in Fig. 1.7(c) becomes a representation of the *total* deformed state of the beam.

It is obvious that any statically determinate beam or frame subjected to proportional loading will have a load–deflection relation generally similar to that of the beam in Fig. 1.7. A plastic hinge will occur at the section which, at any load, has the highest ratio of induced moment to plastic moment, transforming the structure into a mechanism. Thus, the uniform three-pin pitched roof portal frame in Fig. 1.8(a), with load and dimensions as shown, has a bending moment distribution as in Fig. 1.8(b). The maximum moment is 60λ at D, so that, if the plastic moment is 100 units, failure occurs when $60\lambda = 100$, that is, when $\lambda = 1 \cdot 667$. The work equation uses the collapse mechanism in Fig. 1.8(c) in which I_{CD}, the instantaneous centre of CD, lies on AC and ED produced. Suppose that, in this mechanism, ABC rotates ϕ about A, CD rotates θ about I_{CD}, and DE rotates ψ about E. Considering the vertical motion of C,

$$30\phi = 30\theta$$

that is, $\phi = \theta$

2—P.T.S.

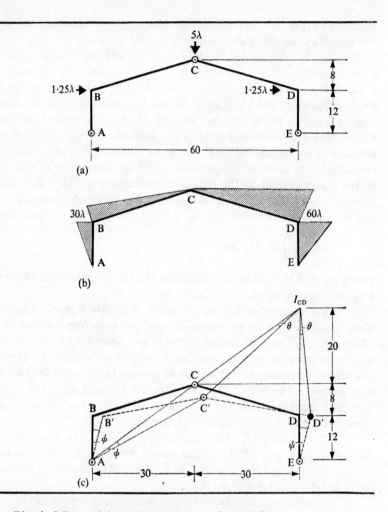

Figure 1.8 Plastic failure of three-pin pitched roof portal frame

Similarly, considering the horizontal motion of D,

$$28\theta = 12\psi$$

that is, $\psi = \frac{7}{3}\theta$

In the work equation, the internal work occurs at the plastic hinge at D where the relative rotation is $(\theta + \psi)$. Hence

$$5\lambda(30\phi) + 1\cdot25\lambda\,(12\phi) + 1\cdot25\lambda\,(12\psi) = 100\,(\theta + \psi) \tag{1.3}$$

giving $\lambda = 1\cdot667$ as before.

1.4   Failure loads of statically indeterminate structures

To illustrate the behaviour of statically indeterminate elastic–plastic structures up to collapse, we take the fixed base portal frame in Fig. 1.9(a), with vertical and horizontal loads remaining proportional and defined in magnitude by the load factor λ. The frame is of uniform section with a plastic moment of $2WL$, and the cross-section is assumed to have unit shape factor.

The load–deflection relations up to collapse with respect to the vertical and horizontal deflections at the mid-span of the beam are shown in Fig. 1.9(b). The end of the elastic phase occurs at a load factor of 1·76, when a plastic hinge forms at E. The structure is capable of sustaining higher loads, and the analysis is continued by treating the frame as elastic except at E, where a hinge rotation at constant moment $2WL$ is allowed to take place. Hinges then form successively at D, A, and C. The structure is then a mechanism [Fig. 1.9(c)], and further deformation takes place at a constant load factor of 2·40. The bending moment distribution at collapse is as shown in Fig. 1.9(d).

A complete elastic–plastic analysis is not essential to a determination of the failure load, provided the correct mechanism can be found. Applying the work equation to the mechanism in Fig. 1.9(c) gives

$$3\lambda W(L\theta) + 2\lambda W(L\theta) = M_p(\theta + 2\theta + 2\theta + \theta) \tag{1.4}$$

that is, $\lambda = 1\cdot20 M_p/WL = 2\cdot40$

It is of interest to try the effect of assuming other mechanisms in the work equation. Thus, the sway mechanism in Fig. 1.9(e) gives

$$3\lambda W(L\theta) = M_p(\theta + \theta + \theta + \theta) \tag{1.5}$$

that is, $\lambda = 2\cdot67$

while the beam mechanism in Fig. 1.9(f) gives

$$2\lambda W(L\theta) = M_p(\theta + 2\theta + \theta) \tag{1.6}$$

that is, $\lambda = 4\cdot00$

These incorrect mechanisms give higher load factors than does the correct mechanism, so that the correct failure load would appear to be derived by considering all mechanisms and using that which gives the least load factor. It is later proved that this is always so—in other words, the plastic work equation applied to an arbitrary mechanism always gives an *upper bound*.

When the shape factor is given a value higher than unity, the formation of each hinge is preceded by the spreading of plastic zones along the member. The corners of the load–deflection curves then become rounded, but the same collapse load is attained.

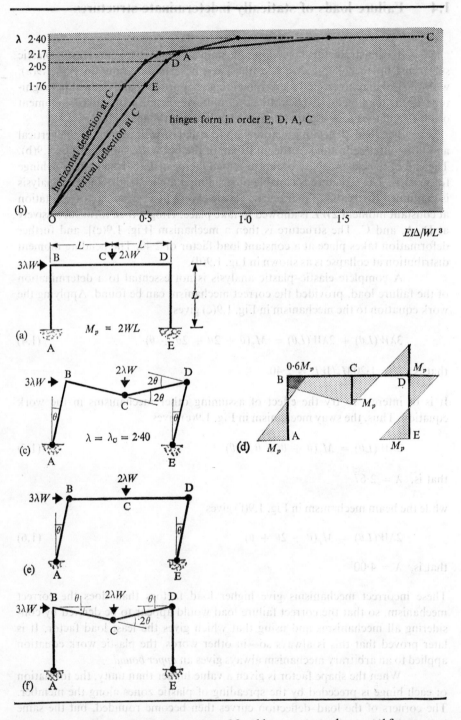

Figure 1.9 Elastic–plastic behaviour of fixed base rectangular portal frame

1.5   Derivation of failure loads by inspection

The failure mechanism for a fixed-ended beam with a uniformly distributed load of intensity w [Fig. 1.10(a)] will evidently be as shown in Fig. 1.10(b), with plastic hinges at ends and centre. The mean descent of the load is $L\theta/4$, whence the work equation becomes

$$wL\left(\frac{L}{4}\theta\right) = M_p(\theta + 2\theta + \theta) \tag{1.7}$$

that is, $w = 16M_p/L^2$

An alternative procedure is to study the bending moment distribution,

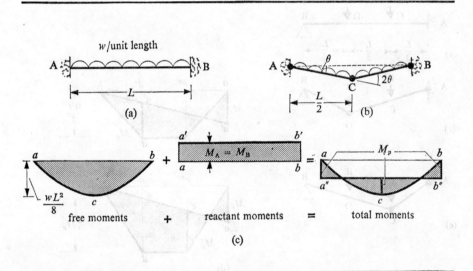

Figure 1.10   Plastic failure of fixed-ended beam with uniformly distributed load

most conveniently by the superimposition of 'free' and 'reactant' moments, as shown in Fig. 1.10(c) (Baker, Horne and Heyman[6]). First, sufficient internal and/or external restraints are removed from the structure to render it statically determinate. In this case, the end fixity is removed, leaving a simply supported beam. The bending moment diagram is then constructed for the resulting statically determinate structure, giving the 'free moments'—in this case, a parabolic distribution of height $wL^2/8$. The reintroduction of the restraints adds the 'reactant moments'—here, a uniform hogging moment of magnitude $M_A = M_B$. The 'total moments' are most conveniently drawn by superimposing a negative version $aa''b''b$ of the reactant moments on the free moment diagram acb, the net moments being the free distribution acb taken to the revised baseline $a''b''$. The collapse solution is achieved by so choosing the reactant moments that the three plastic hinges required to produce a mechanism occur at A, B, and C with numerically equal moments of resistance M_p as shown. It is to be noted that the signs of the plastic moments of resistance (successively hogging, sagging, hogging) agree with the corresponding directions of rotation of the hinges in the plastic hinge mechanism. From Fig. 1.10(c),

$$2M_p = wL^2/8$$

that is, $w = 16M_p/L^2$

as before.

The solution for a propped cantilever with equal loads at the third points, as shown in Fig. 1.11(a), is obtained from the bending moment diagram

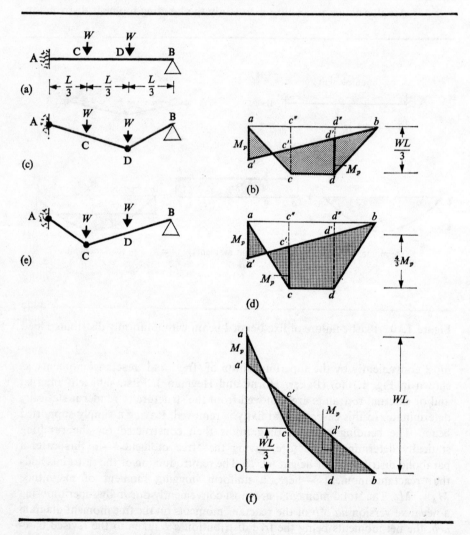

Figure 1.11 Plastic failure of propped cantilever

in Fig. 1.11(b). The free moments *acdb* are obtained by removing fixity at A, and the reactant bending moment line $a'b$ is so drawn that plastic hinges occur at A and D, giving the mechanism in Fig. 1.11(c). Working from the bending moment diagram,

$$\frac{WL}{3} = dd'' = dd' + d'd'' = dd' + \frac{aa'}{3}$$

that is, $\dfrac{WL}{3} = \tfrac{4}{3} M_p$

or $W = 4M_p/L$

It is left to the reader to confirm that the work equation gives the same result.

If the sagging moment is assumed to occur at C instead of at D [Figs 1.11(d) and (e)], then

$$\frac{WL}{3} = cc' + \tfrac{2}{3} aa' = \tfrac{5}{3} M_p$$

or $W = 5M_p/L$

The bending moment diagram immediately shows, however, that this is incorrect, since the bending moment at D is

$$dd' = dd'' - d'd'' = \frac{WL}{3} - \frac{M_p}{3} = \tfrac{4}{3} M_p$$

The particular restraints that are removed to obtain the free moments are not unique. If the support at B in Fig. 1.11(a) is removed, the free moment diagram becomes *acdb* in Fig. 1.11(f). The reactant line $a'b$, representing the effect of the propping force, is so drawn that $a'a = dd' = M_p$, so that

$$WL = Oa = Oa' + a'a = 3dd' + a'a = 4M_p$$

Consideration of these examples shows that a bending moment distribution which represents the collapse conditions must satisfy the following requirements.

(a) *Equilibrium condition.* The bending moments must represent a state of equilibrium between the internal and external loads.

(b) *Mechanism condition.* The plastic moment of resistance must be reached at a sufficient number of sections, and in the necessary senses, for a collapse mechanism to form.

(c) *Yield condition.* The plastic moment of resistance (determined by the value of the yield stress) must nowhere be exceeded.

It will be shown later (*uniqueness theorem*) that a load factor at which a bending moment distribution satisfying these conditions can be found is the unique collapse load factor.

As has already been seen, the use of the mechanism condition in association with a plastic work equation leads to an *upper bound* on the collapse load. Similarly, inspection shows that any bending moment distribution satisfying the equilibrium and yield conditions throughout the structure leads to a *lower bound* on the collapse load. Thus, the fixed-ended beam in Fig. 1.10(a) would develop, in the elastic range, end moments of $wL^2/12$ and a central moment of $wL^2/24$. Hence, the equilibrium and yield conditions will be satisfied by a bending moment distribution which is the same as in the elastic range if $wL^2/12 = M_p$, so that $12M_p/L^2$ is a lower bound on the failure load $(16M_p/L^2)$. The bending moment distribution in Fig. 1.11(d) can be scaled to satisfy the equilibrium and yield conditions for the propped cantilever in Fig. 1.11(a) by multiplying all loads and moments by the factor $\frac{3}{4}$, so that a lower bound on the failure load W becomes $\frac{3}{4}(5M_p/L) = 3 \cdot 75M_p/L$, compared with the actual failure load $4M_p/L$.

1.6   Theorems of plastic collapse

The principles governing plastic collapse which have been justified intuitively will now be stated as formal theorems with proofs (Greenberg,[7] Horne[8]). The remainder of the chapter is then concerned with some consequences of these theorems. The theorems refer to rigid–plastic structures which therefore have zero displacement up to the collapse load. Mechanism deformations are assumed to be infinitely small so that deflections have no effect on the equations of equilibrium. The theorems have relevance, with varying degrees of approximation, to elastic–plastic structures, depending on the influence of change of geometry and instability (Chapter 6).

Reference is made to structures subjected to proportional loading, the intensity of loads being defined by the load factor λ, which is restricted to positive values.

Static or lower bound theorem

If, at any load factor λ, it is possible to find a bending moment distribution in equilibrium with the applied loads and everywhere satisfying the yield condition, then λ is either equal to or less than the load factor at failure.

Hence, any load factor so found is a lower bound on the failure load factor λ_p. The complementary statement may be made that the failure load factor is the greatest at which it is possible to find a bending moment distribution satisfying equilibrium and yield conditions or, expressed succinctly, *the load factor at failure is the greatest static load factor.*

Proof. Let the actual plastic collapse load factor of a structure under loads λW_i be λ_p, and let the collapse mechanism under loads $\lambda_p W_i$ have small hinge rotations θ_j at hinge positions h_j with corresponding plastic moments

M_{pj}. Whatever the sign convention, the sign of any bending moment M_{pj} will be identical with that of the corresponding θ_j. Let the displacements corresponding to the loads $\lambda_p W_i$ in the collapse mechanism be Δ_i. Then

$$\lambda_p \sum W_i \Delta_i = \sum M_{pj} \theta_j \tag{1.8}$$

where, since λ_p and $\sum M_{pj}\theta_j$ are positive, $\sum W_i \Delta_i$ is positive.

Under any load factor λ, let a bending moment distribution be found which satisfies the equilibrium and yield conditions. If the bending moments at the positions h_j are denoted by M_j, then, because of the yield condition,

$$- |M_{pj}| \leqslant M_j \leqslant |M_{pj}| \tag{1.9}$$

Consider now a *virtual work* equation obtained by associating the external loads λW_i (in equilibrium with internal moments M_j at points h_j) with virtual displacements Δ_i (corresponding to internal rotations θ_j at points h_j). Then

$$\lambda \sum W_i \Delta_i = \sum M_j \theta_j \tag{1.10}$$

From Eqns (1.8) and (1.10),

$$(\lambda_p - \lambda) \sum W_i \Delta_i = \sum (M_{pj} - M_j)\, \theta_j \tag{1.11}$$

Since each term $M_{pj}\theta_j$ is positive and in view of Eqn (1.9), each term on the right-hand side of Eqn (1.11) is positive or zero. Hence $\lambda_p \geqslant \lambda$.

Corollary. *The collapse load of a structure cannot be decreased by increasing the strength of any part.*

The bending moment distribution corresponding to the collapse condition of the unmodified structure (load factor λ_{p1}) must satisfy the equilibrium and yield conditions at load factor λ_{p1} for the modified frame (collapse load factor λ_{p2}). Hence $\lambda_{p2} \geqslant \lambda_{p1}$.

Kinematic or upper bound theorem
If, for any assumed plastic mechanism, the external work done by the loads at a positive load factor λ is equal to the internal work at the plastic hinges, then λ is either equal to or greater than the load factor at failure.

Hence, any positive load factor so found is an upper bound on the failure load factor λ_p. The complementary statement is that the failure load factor is the smallest that can be obtained from any arbitrary plastic mechanism by equating internal and external work or, expressed succinctly, *the load factor at failure is the least kinematic load factor.*

Proof. Let any arbitrary mechanism have hinge rotations θ_j at hinge positions h_j, the corresponding plastic moments being M_{pj}. Let the displacements corresponding to loads W_i be Δ_i and let a positive load factor λ be derived from the equation

$$\lambda \sum W_i \Delta_i = \sum M_{pj} \theta_j \tag{1.12}$$

Each term $M_{pj}\theta_j$ is necessarily positive, and, since λ is positive, $\sum W_i\Delta_i$ is positive.

At the actual collapse load factor λ_p for the structure, let the bending moments at the points h_j be denoted by M_j, where the M_j necessarily satisfy the yield condition

$$- |M_{pj}| \leqslant M_j \leqslant |M_{pj}| \tag{1.13}$$

Consider now a *virtual work* equation obtained by associating the external loads $\lambda_p W_i$ (in equilibrium with internal moments M_j at points h_j) with virtual displacements Δ_i (corresponding to internal rotations θ_j at points h_j). Then

$$\lambda_p \sum W_i\Delta_i = \sum M_j\theta_j \tag{1.14}$$

From Eqns (1.12) and (1.14),

$$(\lambda - \lambda_p) \sum W_i\Delta_i = \sum (M_{pj} - M_j)\theta_j \tag{1.15}$$

Since each term $M_{pj}\theta_j$ is positive, and in view of Eqn (1.13), each term on the right-hand side of Eqn (1.15) is positive or zero. Hence $\lambda \geqslant \lambda_p$.

Corollary. *The collapse load of a structure cannot be increased by decreasing the strength of any part.*

The mechanism corresponding to the collapse condition of the unmodified frame (load factor λ_{p1}) provides an upper bound less than or equal to λ_{p1} when used in a work equation for the modified frame (collapse load factor λ_{p2}). Hence $\lambda_{p2} \leqslant \lambda_{p1}$.

It should be noted that the postulation of a plastic mechanism for a structure, leading to a load factor λ, need not necessarily imply that a bending moment distribution in equilibrium with the external loads can exist for such a mechanism. Bending moment distributions do exist (violating the yield condition) for the incorrect mechanisms in Fig. 1.9(e) and (f) for the portal frame in Fig. 1.9(a), and also for the incorrect mechanism in Fig. 1.11(e) for the propped cantilever in Fig. 1.11(a). The correct failure mechanism for the uniform beam AD in Fig. 1.12(a), supporting a uniformly distributed load of intensity w, is as shown in Fig. 1.12(b), the collapse load being $10 \cdot 67 M_p/L^2$. The mechanism in Fig. 1.12(c) gives a valid upper bound of $16M_p/L^2$, but the cantilever ends are not in equilibrium.

Uniqueness theorem
If, at any load factor λ, a bending moment distribution can be found which satisfies the three conditions of equilibrium, mechanism, and yield, then that load factor is the collapse load factor λ_p.

Proof. Since the bending moment distribution satisfies the conditions of equilibrium and yield, $\lambda \leqslant \lambda_p$ by the lower bound theorem. Since the bending moment distribution satisfies the equilibrium and mechanism conditions, the

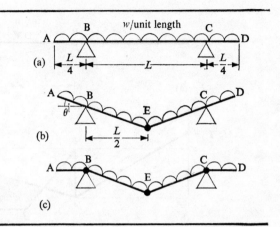

Figure 1.12 (a) Uniformly loaded beam; (b) actual failure mechanism; (c) postulated mechanism, not giving equilibrium state, but giving valid upper bound

plastic work equation obtained by using the mechanism gives the load factor λ, whence $\lambda \geqslant \lambda_p$ by the upper bound theorem. Hence $\lambda = \lambda_p$.

Corollary 1. *The initial internal state of stress has no effect on the collapse load.*

Corollary 2. *If a structure is subjected to any programme of proportional or non-proportional loading, collapse will occur at the first combination of loads for which a bending moment distribution satisfying the conditions of equilibrium, mechanism, and yield can be found.*

These two corollaries follow immediately from the uniqueness theorem, and are of great importance in the application of plastic theory since they show that the collapse load is unaffected by the fabrication or loading history. The second corollary also removes the restricted application of plastic theory to proportional loading, provided all load combinations are examined.

The corollaries apply to elastic–plastic structures provided change of geometry and instability are unimportant, and also provided failure does not occur because of alternating plasticity or incremental collapse. These latter phenomena are discussed in Chapter 5. The collapse loads of elastic–plastic structures are similarly unaffected by thermal stresses, again provided cycles of thermal loading do not cause failure because of alternating plasticity or incremental deformations. Finally, subject to these restrictions, settlement or flexibility of supports and flexibility (as opposed to ultimate strength) of internal connections have no effect on failure loads.

It is important to notice that the uniqueness theorem does *not* state that the bending moment distribution at collapse is itself unique. The continuous beam ABC in Fig. 1.13(a), with uniform plastic moment M_p, has a plastic collapse mechanism in the loaded span with hinges at B and E, the value of W at collapse

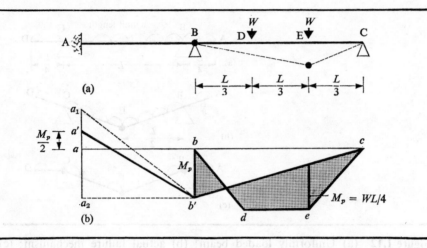

Figure 1.13 Failure of two-span beam with moments within span AB statically indeterminate at collapse

being $4M_p/L$ [compare with Figs 1.11(b) and (c)]. The bending moment distribution in the collapsing span is statically determined [Fig. 1.13(b)], but the reactant line in the span AB can lie anywhere between the limits a_1b' and a_2b', where $a_1a = aa_2 = M_p$. While the bending moments in span AB are *statically indeterminate*, they are not of course indeterminable, provided the complete physical problem is stated. Suppose the beam be elastic–plastic with unit shape factor, the support at A being completely rigid and the simple supports at B and C fixed in position. Let the beam be initially unloaded and free from stress, let loading be proportional and let there be no temperature stresses. Then the bending moment at A when the span BC is on the point of collapse is $M_p/2$ (sagging), giving the reactant line $a'b'$. Where the complete bending moment distribution is not statically determined at collapse, the bending moments depend on such conditions as the initial state of stress, history of loading, settlement of supports, and temperature loading.

1.7   Interaction diagrams

When two independent load systems can act simultaneously on a structure in any ratio, it is useful to study the load factors at collapse by means of an interaction diagram.

The uniform pitched-roof frame in Fig. 1.14(a) has a uniform plastic moment M_p, and, with loads V and H acting as shown, the four mechanisms in Figs 1.14(b)–(e) may be considered. In Fig. 1.14(c), where I_{CD} is the instan-

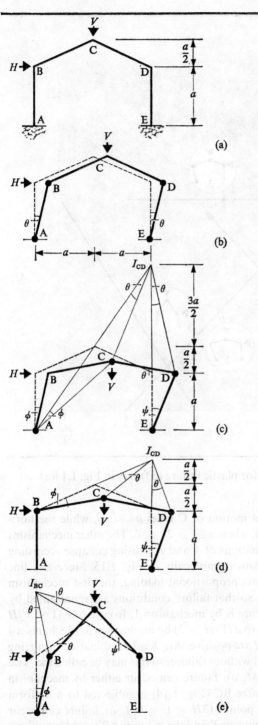

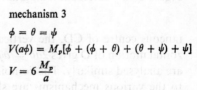

mechanism 1

$$H(a\theta) = M_p(\theta + \theta + \theta + \theta)$$

$$H = 4\frac{M_p}{a}$$

mechanism 2

$$2\phi = 2\theta = \psi$$

$$H(a\theta) + V(a\phi) = M_p[\phi + (\phi + \theta) + \\ (\theta + \psi) + \psi]$$

$$H + V = 8\frac{M_p}{a}$$

mechanism 3

$$\phi = \theta = \psi$$

$$V(a\phi) = M_p[\phi + (\phi + \theta) + (\theta + \psi) + \psi]$$

$$V = 6\frac{M_p}{a}$$

mechanism 4

$$\phi = \theta = \psi$$

$$H(a\phi) - V(a\psi) = M_p[\phi + (\phi + \theta) + \\ (\theta + \psi) + \psi]$$

$$H - V = 6\frac{M_p}{a}$$

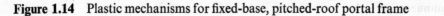

Figure 1.14 Plastic mechanisms for fixed-base, pitched-roof portal frame

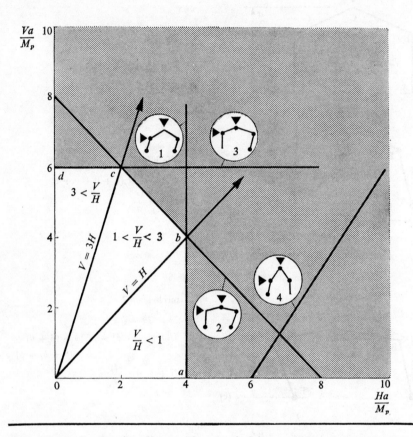

Figure 1.15 Interaction diagram for plastic failure of frame in Fig. 1.14(a)

taneous centre of CD, the vertical motion of C gives $a\phi = a\theta$, while the horizontal motion of D gives $2a\theta = a\psi$, whence $2\phi = 2\theta = \psi$. The other mechanisms are analysed similarly. The combinations of V and H causing collapse according to the various mechanisms are shown graphically in Fig. 1.15. Since any line radiating from the origin represents proportional loading, the first mechanism line intersected represents failure, so that failure conditions are represented by *abcd*. From *a* to *b* ($V/H < 1$), failure is by mechanism 1, from *b* to *c* ($1 < V/H < 3$) by mechanism 2, and from *c* to *d* ($V/H > 3$) by mechanism 3. Mechanism 4 does not occur when both V and H are positive. Any load combination occurring *within* the area O*abcd* is supported without failure, so this may be called the 'safe region'. At point *b* ($H = V = 4M_p/a$), failure can occur either by mechanism 1 or by mechanism 2, and the rafter BC (Fig. 1.14) is subjected to a uniform moment equal to M_p. Similarly at point *c* ($3H = V = 6M_p/a$), failure can occur either by mechanism 2 or by mechanism 3, and the column AB is under uniform moment.

1.8 Number of plastic hinges

In each mechanism in Fig. 1.16, the number of plastic hinges n at collapse is one in excess of the degree of redundancy r of the original structure. This is because, in all these cases, the bending moment distribution becomes statically determinate at failure. The r unknown moments and the value of the load factor at collapse require $r + 1 = n$ plastic hinge moments to enable them to be determined.

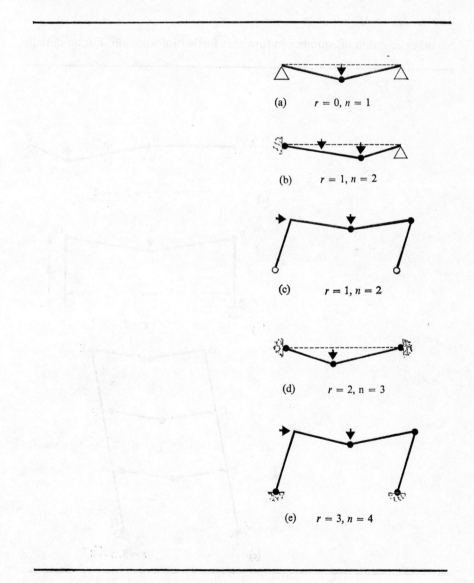

(a) $r = 0, n = 1$

(b) $r = 1, n = 2$

(c) $r = 1, n = 2$

(d) $r = 2, n = 3$

(e) $r = 3, n = 4$

Figure 1.16 Examples of 'complete' collapse

Turning attention to Fig. 1.17, the bending moments are not all determined statically at failure and $n < r + 1$. The beam in Fig. 1.17(a) has already been discussed (Fig. 1.13). In Fig. 1.17(b), the shear forces in the columns must be equal and opposite for horizontal equilibrium, so that if M_A and M_E are the moments at A and E (positive moments producing tension on the inside of the frame),

$$\frac{1}{h}(M_A + M_p) = \frac{1}{h}(M_E + M_p)$$

that is, $M_A = M_E$

No other equation of equilibrium furnishes further information. For an initially

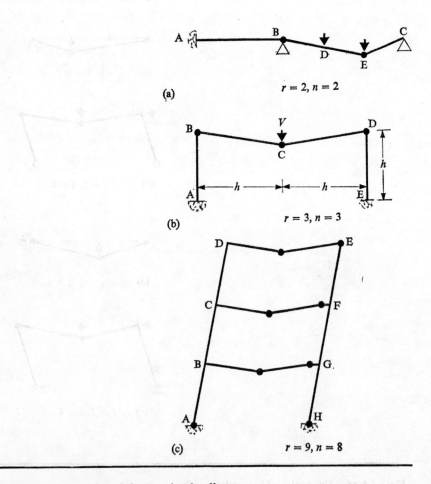

(a) $r = 2, n = 2$

(b) $r = 3, n = 3$

(c) $r = 9, n = 8$

Figure 1.17 Examples of 'incomplete' collapse

stress-free frame of uniform section and unit shape factor, with A and E rigidly fixed, $M_A = M_E = M_p/2$.

The frame in Fig. 1.17(c) has nine degrees of redundancy, since it contains three complete structural rings, and a complete cut in each ring releases three internal forces. Since eight plastic hinges form at collapse, there must be two unknown elements of the bending moment distribution. It is found impossible to determine the column moments at B, C, F, and G, but if one column moment at B or G and one at C or F are defined, the remainder follow from statics. This is discussed further in Chapter 2.

Neal[9] refers to mechanisms of the type shown in Fig. 1.17 as 'partial collapse' mechanisms. Similarly, for particular load combinations at which more than the minimum number of hinges for a mechanism can form (for example, at $V/H = 1$ and 3 for the portal frame in Fig. 1.14), he refers to 'over-complete mechanisms'.

1.9   Collapse load factors in relation to design

It is not the purpose of this volume to discuss in detail the use of plastic collapse load factors in design, but mention must be made of their general significance. A design which is made by specifying that plastic collapse shall not occur at less than a given load factor ('working loads' corresponding to unit load factor) has thereby a known margin of safety with respect to strength, provided any instability effects are suitably taken into account. It may be necessary to consider more than one possible combination of loads or (infrequently) the danger of alternating plasticity or incremental collapse (Chapter 5). It may also be necessary to limit the deflections of the structure at working loads, when it will usually be elastic, and this may lead to a modification of the design. For a wide range of structures it is found, however, that the most important single criterion is the static plastic collapse load. This, together with the fact that the plastic collapse load is relatively easy to calculate and that minimum weight theorems can be derived (Chapter 4), encourages the use of plastic theory as the basis of many design procedures.

Bibliography

1   Horne, M. R. Elastic–plastic failure loads of plane frames. *Proc. R. Soc.*, 1963, **274** [A], 343.

2   Horne, M. R. and W. F. Merchant. *The Stability of Frames*, Pergamon Press, Oxford, 1965.

3   Smith, J. O. and O. M. Sidebottom. *Inelastic Behaviour of Load-Carrying Members*, John Wiley, New York, 1965.

4   Horne, M. R. The effect of strain-hardening on the equalisation of moments in the simple plastic theory. *Welding Research*, 1951, **5**, 147.

5   Hrennikoff, A. Importance of strain-hardening in plastic design. *J. Struct. Div. Am. Soc. civ. Engrs*, 1965, **94** (ST4) 23.

6   Baker, J. F., M. R. Horne, and J. Heyman. *The Steel Skeleton*, Vol. II, Cambridge University Press, London, 1956.

7   Greenberg, H. J. The principle of limiting stress for structures, Second Symposium on Plasticity, Brown University, 1949.

8   Horne, M. R. Fundamental propositions in the plastic theory of structures. *J. Instn civ. Engrs*, 1950, **34**, 174.

9   Neal, B. G. *The Plastic Methods of Structural Analysis*, Chapman and Hall, London, 1956.

Problems

1.1   Determine the plastic moduli and shape factors of the following cross sections:

(a)   An I-section of total depth d and flange width $0 \cdot 5d$, the web and flange thicknesses being $0 \cdot 1d$, bent about the major axis;

(b)   as (a), bent about the minor axis;

(c)   solid circular bar of diameter d;

(d)   thin circular tube of diameter d and wall thickness t;

(e)   solid square bar of side d, bent about a diagonal;

(f)   thin square tube of side d and wall thickness t, bent about an axis perpendicular to a side;

(g)   as (f), bent about a diagonal.

1.2   The continuous beam ABCD in Fig. E1.1 has a uniform section. If collapse just occurs under loads as shown, determine the value of the plastic moment;

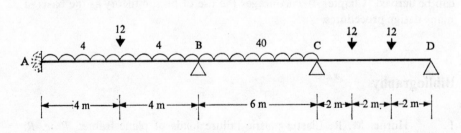

total loads in kilonewtons, dimensions in metres.

Figure E1.1

(a)   by considering all reasonable mechanisms;
(b)   graphically.

1.3   A member AB, with uniform plastic moment M_p and of length L, is fixed in position and direction at A and simply supported at B. If it carries a distributed load the intensity of which varies linearly from w at A to zero at B, find the position of the sagging plastic hinge and the value of w at collapse.

1.4   A uniform beam of length L, simply supported at its ends, is subjected to a uniformly distributed load of intensity w. If the plastic moment is M_p, what is the maximum possible value of w at collapse:
(a)   if the beam is propped at one place within its span;
(b)   if the beam has n props?
      In case (b), determine the spacing of the props.

1.5   A cantilever AB, rigidly fixed at A, is of length L and has a uniform plastic moment M_p. It carries two equal loads W, one at B and the other at the centre C. The cantilever is simply propped at D, distance x from end B. Calculate the value of x which will make W a maximum at collapse, and the corresponding value of W.

1.6   The projecting corner ABC of a floor system, shown in plan in Fig. E1.2, consists of continuous, rigidly connected, equally spaced beams of equal cross section, rigidly fixed along AB. Ignoring the resistance of the beams to twisting, calculate the value of the vertical concentrated load at C which would cause collapse if the plastic moment of the beams about horizontal axes is M_p.

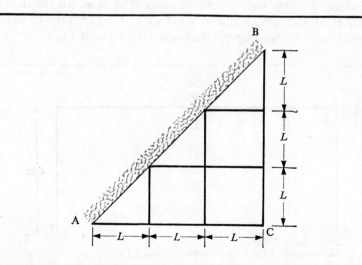

Figure E1.2

1.7   The long continuous member AB as shown in plan in Fig. E1.3 is
       hinged to a number of equally spaced horizontal cantilevers of length
       2L, each of plastic moment M. The member AB has a plastic moment
       about a horizontal axis of 2M, while the cantilevers are spaced L
       apart. Show that, if a vertical point load W just sufficient to cause
       collapse is placed on AB at the end of one of the cantilevers, plastic
       hinges will form in seven cantilevers and $W = 4M/L$.

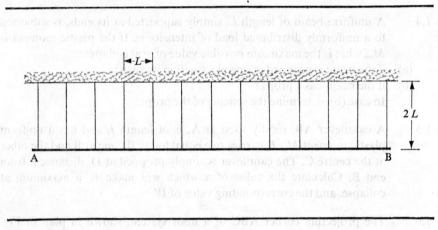

Figure E1.3

1.8   The cantilevers in Example 1.7 are made less strong and more numerous,
       being equally spaced at an interval which is small compared with their
       length 2L. If the cantilevers now develop a plastic moment of m per unit
       distance along the wall, the plastic moment of beam AB remaining at
       2M, show that the maximum point load W which may be applied
       anywhere within the length of the beam AB is $4\sqrt{(mM/L)}$.

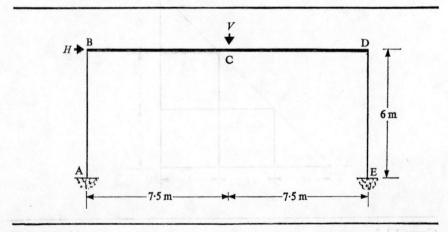

Figure E1.4

1.9 The portal frame in Fig. E1.4 is fixed at A and E and carries loads V and H as shown. The beam has a plastic moment of 100 kNm and the columns have a plastic moment of 60 kNm. Show on a graph of V versus H the load combinations that are just sufficient to cause plastic collapse.

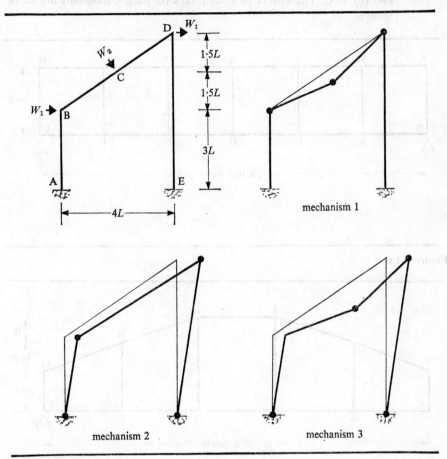

Figure E1.5

1.10 Consider the mechanisms 1, 2, and 3 for the loaded frame in Fig. E1.5, the plastic moment being of uniform value M_p. Show by drawing the interaction diagram that, for all positive combinations of W_1 and W_2, mechanism 1 never applies. Derive the particular solution for $W_2 = 10W_1$ and, using a virtual work equation, find the bending moment at B in terms of the plastic moment M_p.

1.11 The simply supported Vierendeel girder in Fig. E1.6 has upper and lower chord members with plastic moments of 150 kip ft and vertical members with plastic moments of 90 kip ft. Find the single concentrated vertical load that will cause collapse when applied (a) at D, (b) at C, and (c) at B. Ignore the effects of axial forces on plastic moments.

1.12   All the members of the simply supported Vierendeel girder in Fig. E1.7 have a plastic moment of 300 kip ft. Find the values of a single concentrated load that will just cause plastic collapse when applied (a) at B and (b) at C. The effects of axial forces on plastic moments are to be ignored.

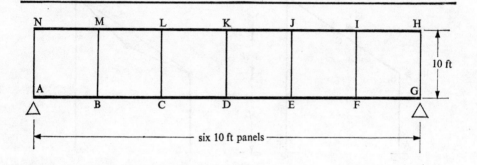

Figure E1.6

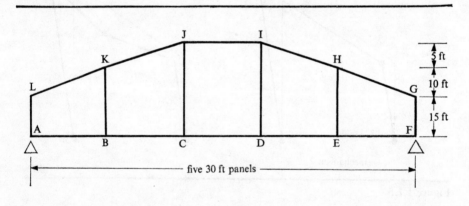

Figure E1.7

Chapter 2
Methods of plastic analysis

2.1   Introduction

The plastic collapse loads of very simple structures, such as those considered in
Chapter 1, may be derived by inspection. In effect, the *uniqueness theorem* is
being used by ensuring the simultaneous satisfaction of the three conditions of
equilibrium, mechanism, and yield (Section 1.5). In simple structures in which
it is not clear where hinges will form, because of the presence of many loading
points, the use of free and reactant bending moment diagrams is of considerable
assistance. This method has been extensively applied to the analysis of single
bay bents. A typical example of such an application is discussed in Section 2.2.
The extension of the method to multi-bay bents is described in Section 2.3.

In analysing structures with larger numbers of members and joints, it is
found to be easier to invoke *either the lower or the upper bound theorem*. The
lower bound theorem is used in the method of inequalities (Neal and Symonds[1]),
in which the yield condition is satisfied for each section at which a plastic hinge
may occur by writing the continuing inequality $-M_p \leqslant M \leqslant M_p$. A certain
number of independent equations relating the bending moments M to the load
factor λ are derived from the equilibrium conditions, and the collapse load
factor is the minimum value of λ at which the inequalities can be satisfied subject
to these equilibrium conditions. The method is arithmetically tedious, but is
systematic and is a convenient basis for the derivation of collapse loads by digital
computer. It is described at the end of the chapter in Section 2.7.

The kinematic or upper bound theorem provides the most convenient
hand methods of deriving failure loads. Since collapse mechanisms are easily
visualized, difficulties over sign conventions are avoided, and intuitive ideas of
structural behaviour are readily introduced. The exploration of all possible
mechanisms becomes tedious for any but the simplest structures, but by exploring so-called *elementary mechanisms* some idea of the general nature of the
failure mechanism may be obtained. In the method of *combined mechanisms*
described in Section 2.4, a specified number of elementary mechanisms are first
considered, and these are then combined in such a way as to decrease the
load factor. A final static check is made to determine whether the combined

mechanism which appears to have the lowest load factor satisfies the yield condition; that is, the uniqueness theorem is finally invoked to ensure that a valid failure mechanism has been postulated. A general method of carrying out a static check is described in Section 2.5.

The presence of distributed loads introduces some difficulty when using mechanism methods for extensive frames, since sagging plastic hinges in beams will usually occur at initially unknown sections away from mid-span. This is discussed in Section 2.6.

2.2   Graphical procedure for portal frames

Frames with fixed bases

In a single bay bent such as the fixed-base, pitched-roof portal frame in Fig. 2.1(a), the loads are usually applied at discrete points by secondary members such as purlins and sheeting rails. Any uniformly distributed loads (such as those due to self-weight) may also for convenience be considered as concentrated at these positions, and a typical total loading condition (due to vertical loads plus wind) is as shown. The loads are working values (in kilonewtons, or kN) and the distances are given in metres. It is desired to design the frame for static collapse at a load factor of not less than 1·60, the yield stress being 250 N/mm². The calculations will be performed for unit load factor, and the final required plastic moment scaled to give the requisite load factor.

In using the free and reactant diagram method (Section 1.5), it is first necessary to render the frame statically determinate. This is most conveniently achieved in this case by introducing a complete cut at the apex I, the internal forces thereby removed being represented [Fig. 2.1(b)] by a bending moment M, a horizontal thrust H, and a vertical force V. The *free moments* (producing tension on the outside of the frame) are plotted to the base-line AQ in Fig. 2.1(c) as $acIoq$, the numerical values also being shown. The *reactant moments* (producing compression on the outside) are plotted to the same baseline as $a_1c_1i_1o_1q_1$, the leading moments being given in terms of M, H, and V by

$$\left.\begin{aligned}
Aa_1 &= M + 7H - 9V \\
Cc_1 &= M + 3H - 9V \\
Ii_1 &= M \\
Oo_1 &= M + 3H + 9V \\
Qq_1 &= M + 7H + 9V
\end{aligned}\right\} \tag{2.1}$$

The *resultant moments* are represented by the shaded area, these moments being given by the free moment line $acIoq$ read to the reactant moment line $a_1c_1i_1o_1q_1$ as base.

It is impossible merely by inspection to determine what values of the three unknown internal reactions M, H, and V and plastic moment M_p will just produce a collapse mechanism, so the next step is to postulate a likely mechanism. That in Fig. 2.1(d) is chosen, giving plastic hinge moments at A, I, O, and Q of values M_p, $-M_p$, M_p, and $-M_p$ respectively, where positive moments cause tension outside. Hence from Fig. 2.1(c), using Eqns (2.1),

$$
\left.
\begin{aligned}
a_1a &= 352 - M - 7H + 9V = M_p \\
i_1I &= 0 - M = -M_p \\
o_1o &= 180 - M - 3H - 9V = M_p \\
q_1q &= 44 - M - 7H - 9V = -M_p
\end{aligned}
\right\} \tag{2.2}
$$

These equations, one for each of the four plastic hinges, suffice to determine the three unknown internal reactions M, H, and V (*three* degrees of redundancy) and the unknown value of M_p. It is found that

$$
\begin{aligned}
M_p &= M = 96\cdot9 \text{ kNm} \\
H &= 14\cdot44 \text{ kN}
\end{aligned}
$$
and $\quad V = -6\cdot23 \text{ kN}$

The reactant line in Fig. 2.1(c) is that derived from these values. Inspection shows that the resultant moments are in excess of M_p at B, C, G, and H, so that the incorrect mechanism has been assumed and a second trial must be made.

In choosing a revised mechanism, guidance is obtained from the resultant bending moment diagram in Fig. 2.1(c). It is evident that a positive plastic hinge (tension outside) will be formed at C rather than at A, and that a negative plastic hinge will be formed at either G or H rather than at I. The moments at G and H are

$$
\begin{aligned}
M_G &= 40 - M - H + 3V = -90\cdot0 \text{ kN} \\
M_H &= 10 - M - 0\cdot5H + 1\cdot5V = -103\cdot4 \text{ kN}
\end{aligned}
$$

Hence hinges are to be expected at C, H, O, and Q, giving the mechanism depicted in Fig. 2.1(e). The new position of the reactant line $a_2c_2i_2o_2q_2$ in Fig. 2.1(f) is derived from the four equations

$$
\left.
\begin{aligned}
c_2c &= 360 - M - 3H + 9V = M_p \\
h_2H &= 10 - M - 0\cdot5H + 1\cdot5V = -M_p \\
o_2o &= 180 - M - 3H - 9V = M_p \\
q_2q &= 44 - M - 7H - 9V = -M_p
\end{aligned}
\right\} \tag{2.3}
$$

whence $\quad M_p = 110\cdot8 \text{ kNm}$
$\qquad\ \ M = 95\cdot0 \text{ kNm}$
$\qquad\ \ H = 21\cdot4 \text{ kN}$
$\qquad\ \ V = -10\cdot0 \text{ kN}$

It is readily ascertained that, with the reactant line constructed according to these values [as in Fig. 2.1(f)] the plastic moment is nowhere exceeded, and the value of M_p for unit load factor is thus as derived. For a load factor of $1\cdot60$ and with a yield stress of 250 N/mm², the required plastic modulus Z_p is therefore

$$\frac{(110.8 \times 10^6) \times 1.6}{250} \text{ mm}^3 = 709 \text{ cm}^3$$

A Universal Beam of serial size $16 \times 5\frac{1}{2}$ in at 26 lb/ft has $Z_p = 719 \text{ cm}^3$, and would therefore be satisfactory.

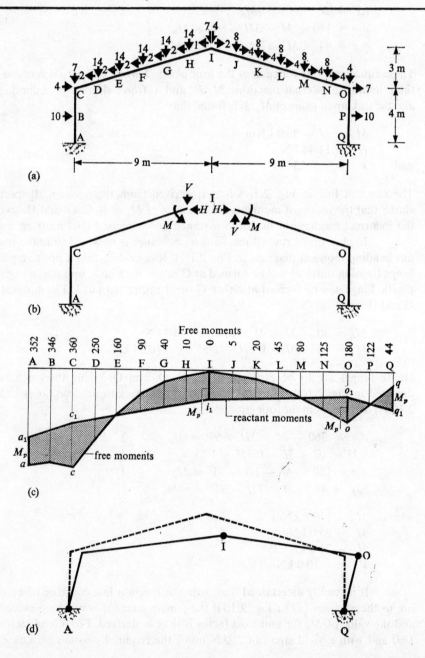

(a)

(b)

(c)

(d)

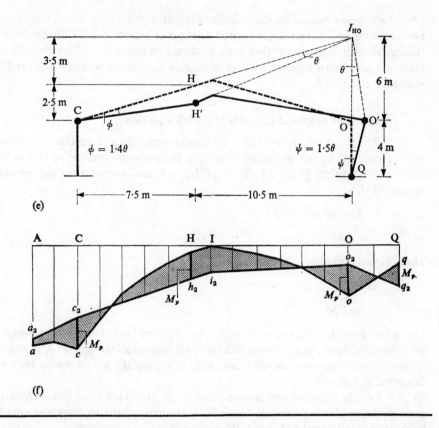

Figure 2.1 (a) Pitched roof frame with loading; (b) Cut at apex with redundant forces; (c) Graphical solution for plastic moments at A, I, O, and Q; (d) Mechanism corresponding to (c); (e) Revised mechanism corresponding to (f); (f) Graphical solution for plastic moments at C, H, O, and Q.

It is interesting to note that the solution for the incorrect mechanism represented by Fig. 2.1(c) not only shows that the required M_p at unit load factor has a lower bound of 96·9 kNm, but also provides an upper bound. The maximum resultant bending moment occurs at C and is

$$M_C = 360 - M - 3H + 9V = 163·7 \text{ kNm}$$

With a plastic moment of this value, the resultant bending moments in Fig. 2.1(c) would represent a distribution satisfying the equilibrium and yield conditions so that, by the lower bound theorem, the collapse load would be greater than or equal to unity. Hence 163·7 kNm represents an upper bound on the plastic moment required at unit load factor, and from Fig. 2.1(c) it is therefore deduced that

$$96·9 \leqslant M_p \leqslant 163·7 \text{ kNm}$$

It should be clearly understood that the 'static or lower bound' theorem provides a lower bound on the collapse load of a given structure, *or* an upper bound on the plastic moment required to support a given load. Conversely, the 'kinematic or upper bound' theorem establishes an upper bound on the collapse load of a given structure, *or* a lower bound on the plastic moment required to support a given load.

Check of graphical procedure by work equation

The value of M_p obtained by considering free and reactant moments may be checked by the work equation. The instantaneous centre of HO in the collapse mechanism [Fig. 2.1(e)], is at I_{HO} whence, considering the vertical motion of H,

$$7 \cdot 5\phi = 10 \cdot 5\theta$$

or $\qquad \phi = 1 \cdot 4\theta$

The horizontal motion of O gives

$$4\psi = 6\theta$$

or $\qquad \psi = 1 \cdot 5\theta$

The work done by the external loads may be obtained directly by summing the individual items, but a convenient device is to use instead *any set of moments in equilibrium with the external loads*. This is a general method which may be described as follows.

Let the collapse mechanism have hinge rotations θ_j at points h_j with corresponding plastic moments M_{pj}, and let the accompanying displacements Δ_i be associated with applied loads $\lambda_p W_i$, so that

$$\lambda_p \sum W_i \Delta_i = \sum M_{pj} \theta_j \qquad (2.4)$$

Let any bending moment distribution in equilibrium with applied loads W_i have bending moments M_j at points h_j. Then by applying a virtual work equation using rotations θ_j and displacements Δ_i,

$$\sum W_i \Delta_i = \sum M_j \theta_j \qquad (2.5)$$

Hence, from Eqns (2.4) and (2.5),

$$\lambda_p \sum M_j \theta_j = \sum M_{pj} \theta_j \qquad (2.6)$$

Now the free moments $ac1oq$ to base AQ in Fig. 2.1(c) are a set of moments in equilibrium with the loads in Fig. 2.1(a) at unit load factor. Hence these free moments may be used for the moments M_j in Eqn (2.6), so that with rotations θ_j from the collapse mechanism in Fig. 2.1(e), the plastic moment required when $\lambda_p = 1 \cdot 00$ is given by

$$360\phi - 10(\phi + \theta) + 180(\theta + \psi) - 44\psi$$
$$= M_p\{\phi + (\phi + \theta) + (\theta + \psi) + \psi\} \qquad (2.7)$$

should be noted that, on the left-hand side of Eqn (2.7), proper account is taken of the signs of the hinge rotations in relation to the free moments. Substituting for ϕ and ψ in terms of θ,

$$(360 \times 1\cdot4 - 10 \times 2\cdot4 + 180 \times 2\cdot5 - 44 \times 1\cdot5)\theta$$
$$= M_p(1\cdot4 + 2\cdot4 + 2\cdot5 + 1\cdot5)\theta \quad (2.8)$$

Hence $M_p = 110\cdot8$ kNm as before.

Frames with pinned bases

The frame in Fig. 2.1 but with pinned bases may be analysed by using the same free moments, on condition that the three internal reactions M, H, and V at I [Fig. 2.1(b)] are so chosen that no bending moments result at A and Q. This requires that

$$\left. \begin{array}{l} 352 - M - 7H + 9V = 0 \\ 44 - M - 7H - 9V = 0 \end{array} \right\} \quad (2.9)$$

Whatever collapse mechanism is assumed, these conditions must be satisfied, so that the three 'unknowns' M, H, and V are effectively reduced to one. The failure mechanism is found to be that shown in Fig. 2.2(a), whence the resultant bending moment diagram in Fig. 2.2(b) also satisfies the condition that the plastic moment should be reached at H and O. Hence.

$$\left. \begin{array}{l} h_1H = 10 - M - 0\cdot5H + 1\cdot5V = -M_p \\ o_1o = 180 - M - 3H - 9V = M_p \end{array} \right\} \quad (2.10)$$

The solution to Eqns (2.9) and (2.10) is

$$M_p = 165\cdot5 \text{ kNm}$$
$$M = 146\cdot0 \text{ kNm}$$
$$H = 7\cdot5 \text{ kN}$$
$$V = -17\cdot1 \text{ kN}$$

A check by the work equation may again be made by using the collapse mechanism in Fig. 2.2(a) in association with the free moments as a set of moments in equilibrium with the applied loads. In the collapse mechanism, the vertical motion of H gives $\phi = 1\cdot4\theta$ as before, and the horizontal motion of O gives

$$4\psi = 11\cdot6\theta$$

or $\quad \psi = 2\cdot9\theta$

Hence, using Eqn (2.6),

$$(352 \times 1\cdot4 - 10 \times 2\cdot4 + 180 \times 3\cdot9 - 44 \times 2\cdot9)\theta$$
$$= M_p(2\cdot4 + 3\cdot9)\theta \quad (2.11)$$

giving $M_p = 165\cdot5$ kNm as before. Note that, on the right-hand side of Eqn (2.11), the 'plastic moments' of the hinges at the base are regarded as zero, thus allowing for the fact that structural hinges are assumed at these sections.

At a load factor of 1·60, the Z_p required for a plastic moment of 165·5 kNm is 1060 cm³, which is provided by a Universal Column of serial size 16 × 7 in at 40 lb/ft (Z_p = 1191 cm³).

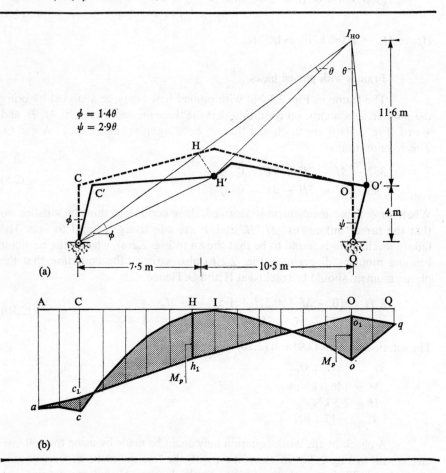

Figure 2.2   (a) Mechanism for frame with pinned bases; (b) Graphical solution

2.3   Multi-bay frames

In multi-bay pitched roof frames, vertical loading is usually more critical for design than the loading conditions imposed by wind loads. When the same cross section is used throughout a frame composed of equally dimensioned bays, it is only necessary to consider the design of the two outermost bays, since the bending moment distribution for an outer bay at the point of collapse gives immediately a distribution for the internal bays satisfying, for these bays, the equilibrium and yield conditions. This may be illustrated by considering the multi-bay frame in Fig. 2.3.

The dimensions (in feet) and the working loads (in kips) are as shown in Fig. 2.3(a), the frame being assumed to be five bays wide. Free moments for the first bay ABFJ are obtained by making a cut at the apex F, giving for ABF the free moments *abcde*F in Fig. 2.3(d). The free moments for the roof members FIJ are symmetrical with those for FCB. The disposition of the plastic hinges at collapse [see Fig. 2.3(c)] is also symmetrical, whence the reactant line for the roof members must also be symmetrical. It is therefore sufficient to consider only the reactant line $a_1b_1c_1d_1e_1f_1$ in Fig. 2.3(d). Because of the symmetry, the vertical internal force V_1 in Fig. 2.3(b) must be zero, leaving only a horizontal thrust H_1 and an apex bending moment M_1. Hence from Fig. 2.3(d), there are three equations, one each for plastic hinges at A, B, and E, as follows,

$$\left.\begin{aligned}
a_1a &= 320 - M_1 - 30H_1 = -M_{p1} \\
b_1b &= 320 - M_1 - 15H_1 = M_{p1} \\
e_1e &= 20 - M_1 - 2{\cdot}5H_1 = -M_{p1}
\end{aligned}\right\} \tag{2.12}$$

where M_{p1} is the required plastic moment. Hence $M_{p1} = 81{\cdot}9$ kip ft, $H_1 = 10{\cdot}9$ kips and $M_1 = 74{\cdot}7$ kip ft.

Consider now the second bay. If the first bay stood alone, the entire equilibrium moments would be obtained by adding to the bending moment diagram in Fig. 2.3(d) an image of itself reflected about F. A set of moments satisfying yield and equilibrium for the second bay would be a repeat of this complete bending moment diagram, and so on for the remaining bays. Figure 2.3(e) shows the complete bending moment diagram. It should be noted that the bending moments in the internal columns exactly cancel. Since the resulting bending moment diagram represents the absence of plastic hinges in the internal columns, failure is confined to the outer bays, and so the collapse mode for the entire frame will be as shown in Fig. 2.3(c).

The internal bays do not collapse because they are being reacted upon by the outermost bays. If desired, the section of the members KJOS in Fig. 2.3(a) may be reduced below the value for the outer bay. This possibility may be investigated by the graphical method as follows.

The collapsing external bay applies to the joint at J in Fig. 2.3(c) an anticlockwise moment of $M_{p1} = 81{\cdot}9$ kip ft, together with a horizontal thrust of $H_1 = 10{\cdot}9$ kips as shown in Fig. 2.4(a). When these reactions are added to the vertical loads acting in the second bay the free moments for KJO become $kjj'lmn$O in Fig. 2.4(c). The two values at J (238·1 and 320 kip ft) refer to the column and roof at J respectively. Since the collapse mechanism [Fig. 2.4(b)] is again found to be symmetrical with respect to plastic hinges in the roof members, the internal reaction V_2 [Fig. 2.4(a)] is zero, and three equations are sufficient to obtain a solution. If M_{p2} is the plastic moment of members KJLORS, the equations become as follows for hinges in the column at K and in the roof members at J and N.

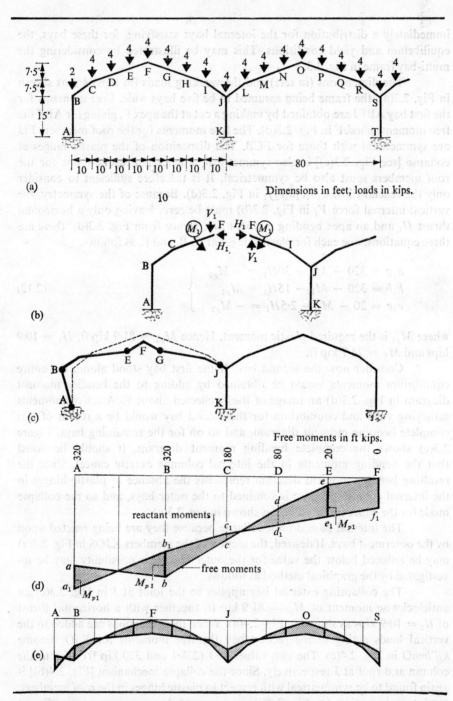

Figure 2.3 (a) Multi-bay frame (five bays); (b) Cut at apex of first bay, with redundant forces; (c) Failure mechanism for end bay; (d) Graphical solution for failure in end bay (left-hand rafter only); (e) Bending moment diagram for whole frame satisfying equilibrium and yield conditions

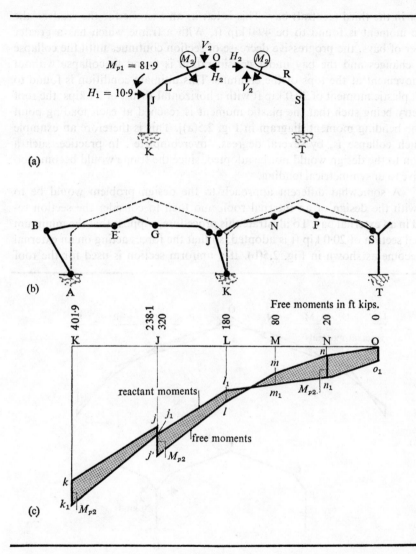

Figure 2.4 (a) Cut at apex of second bay, with internal forces and forces from end bay; (b) Failure of two end bays (reduced section in second bay); (c) Graphical construction for failure in second bay (left-hand rafter only)

$$\left.\begin{array}{l} k_1k = 401\cdot9 - M_2 - 30H_2 = -M_{p2} \\ j_1j' = 320 - M_2 - 15H_2 = M_{p2} \\ n_1n = 20 - M_2 - 2\cdot5H_2 = -M_{p2} \end{array}\right\} \tag{2.13}$$

The solution of these equations gives $M_{p2} = 63\cdot0$ kip ft, $H_2 = 13\cdot9$ kips, and $M_2 = 48\cdot2$ kip ft.

Since the horizontal thrust has increased from 10·9 kips in the first bay to 13·9 kips in the second, the section of the members could be reduced even

4—P.T.S.

further in the third (central) bay. This is left as an exercise for the reader—the plastic moment is found to be 49·0 kip ft. With a frame which has a greater number of bays, the progressive decrease of section continues until the collapse mode changes and the bay under consideration is found to collapse without any movement at the tops of the columns. This ultimate condition is found to yield a plastic moment of 20·0 kip ft with a horizontal thrust of 24 kips, the roof geometry being such that the plastic moment is reached at each loading point [see the bending moment diagram in Fig. 2.5(a)]. This is therefore an example in which collapse is, by several degrees, 'overcomplete'. In practice, such a solution to the design would not be adopted, since the frame would become too sensitive to unsymmetrical loading.

A somewhat different approach to the design problem would be to start with the design of an internal roof, and then to consider the section required in an external bay. To illustrate this procedure, suppose that the minimum internal section of 20·0 kip ft is adopted, so that the forces acting on an external bay become as shown in Fig. 2.5(b). If a uniform section is used for the roof

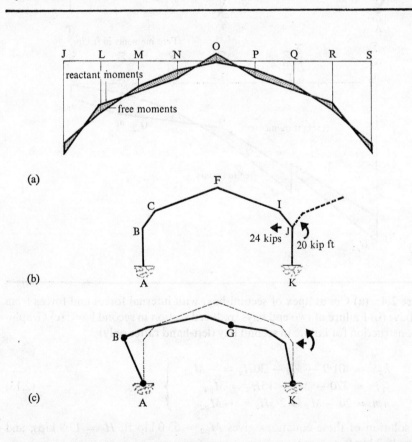

Figure 2.5 (a) Graphical solution for minimum section in three central bays; (b) Forces acting on external bay; (c) Failure of external bay

members and both columns in the external bay, the collapse mode is as shown in Fig. 2.5(c) and the plastic moment becomes 109·5 kip ft. (It should be noted that the effects of axial load have not been allowed for in the above analysis. They would be of importance in the internal bays because of the low plastic moment and simultaneous high thrust.)

This example shows the adaptability of the plastic approach to design. Whereas elastic design methods have to allow for the effect of the flexural rigidity of one part of a structure on stresses induced in the rest of the structure, the reactions exerted by one part of a structure on another at the collapse condition are much more under the control of the designer. The design procedure then becomes direct, in contrast to indirect elastic methods in which assumptions have to be made about relative stiffness ratios before any analysis can be carried out.

2.4   Method of combined mechanisms

This method (Neal and Symonds[2]) is most suitable for rigid frames containing horizontal and vertical members only, although it may be used for structures with sloping members. It will be introduced by considering the fixed-base two-bay frame in Fig. 2.6(a) with dimensions, loads, and plastic moments as shown. The aim is to explore the mechanisms most likely to cause collapse, using the upper bound theorem, according to which the lowest load factor obtained from all possible mechanisms is the correct value.

In exploring possible mechanisms, it is first noted that plastic hinges will only occur under concentrated loads or at joints, since bending moments must vary linearly between such sections, whence ten possible hinge positions exist, as shown in Fig. 2.6(b). Hinges are shown at the tops of the outer columns but not at the outer ends of the beams, because the columns have a smaller plastic moment than either of the beams. The three *elementary mechanisms* in Figs 2.6(c), (d), and (e) are first considered, and work equations written as follows.

Sway mechanism [Fig. 2.6(c)]

$$\left.\begin{aligned}
4\lambda(15\theta) &= 20(\theta + \theta + \theta + \theta + \theta + \theta) \\
60\lambda\theta &= 120\theta \\
\lambda &= 2\cdot000
\end{aligned}\right\} \tag{2.14}$$

Beam mechanism BD [Fig. 2.6(d)]

$$\left.\begin{aligned}
6\lambda(10\theta) &= 20(\theta) + 50(2\theta + \theta) \\
60\lambda\theta &= 170\theta \\
\lambda &= 2\cdot833
\end{aligned}\right\} \tag{2.15}$$

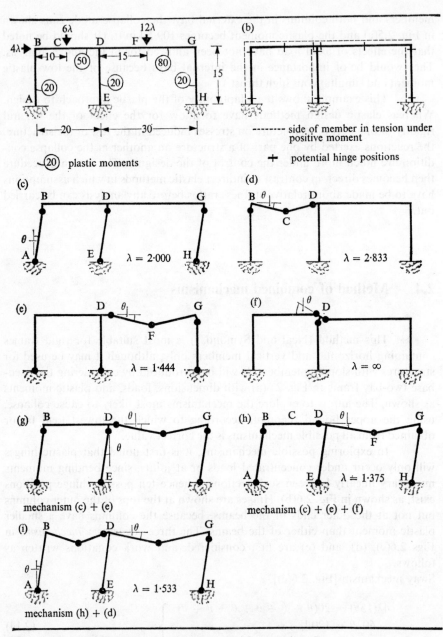

side of member in tension under positive moment

⌒20 plastic moments

+ potential hinge positions

$\lambda = 2 \cdot 000$

$\lambda = 2 \cdot 833$

$\lambda = 1 \cdot 444$

$\lambda = \infty$

mechanism (c) + (e)

mechanism (c) + (e) + (f) $\lambda \doteq 1 \cdot 375$

mechanism (h) + (d) $\lambda = 1 \cdot 533$

Figure 2.6 Combined mechanisms method for two-bay frame

Beam mechanism DG [Fig. 2.6(e)]

$$
\left.
\begin{aligned}
12\lambda(15\theta) &= 80(\theta + 2\theta) + 20(\theta) \\
180\lambda\theta &= 260\theta \\
\lambda &= 1\cdot444
\end{aligned}
\right\}
\tag{2.16}
$$

Restricting attention to these mechanisms, the lowest load factor is 1·444, whence $\lambda_p \leqslant 1\cdot444$. These elementary mechanisms will now be combined to see whether a lower load factor is obtainable. Since none of the three mechanisms in Figs 2.6(c), (d), or (e) involves a possible rotation of the central joint D, it is necessary to add the rotation of this joint [Fig. 2.6(f)] as a fourth elementary mechanism (for which the load factor is infinite).

The obvious step is to combine the two elementary mechanisms with the lowest load factors, namely the sway mechanism for which $\lambda = 2\cdot000$ and the beam mechanism DG for which $\lambda = 1\cdot444$, giving the mechanism in Fig. 2.6(g). When this is done, however, the joint D cannot remain in equilibrium, since the clockwise moments acting on the joint from beam DG and column DE total $80 + 20 = 100$ units, whereas the plastic moment of beam BD is only 50 units. This indicates that the joint at D must rotate, that is, the joint mechanism in Fig 2.6(f) is also added, giving the mechanism in Fig. 2.6(h). The resultant work equation is obtained by adding the work Eqns (2.14) and (2.16) for the elementary mechanisms, and making adjustments because of the rotation of the central joint. Hence

$$\left.\begin{array}{c} (60 + 180)\lambda\theta = (120 + 260)\theta - 80\theta - 20\theta + 50\theta \\ 240\lambda\theta = 330\theta \\ \lambda = 1\cdot375 \end{array}\right\} \tag{2.17}$$

The terms 80θ and 20θ are subtracted on the right-hand side to cancel the plastic hinges which disappear at D in beam DG and column DE in proceeding from Fig. 2.6(g) to Fig. 2.6(h), and the term 50θ is added for the plastic hinge introduced at the end of beam BD. Hence $\lambda_p \leqslant 1\cdot375$.

The effect may be explored of adding the remaining elementary mechanism [Fig. 2.6(d)] to the combined mechanism in Fig. 2.6(h), thus giving the combined mechanism in Fig. 2.6(i). Since two opposite plastic hinge rotations are cancelled at B the combined work equation [obtained from Eqns (2.15) and (2.17) becomes

$$\left.\begin{array}{c} (60 + 240)\lambda\theta = (170 + 330)\theta - 2 \times 20\theta \\ 300\lambda\theta = 460\theta \\ \lambda = 1\cdot533 \end{array}\right\} \tag{2.18}$$

Since the combined mechanism in Fig. 2.6(i) gives a higher load factor than that in Fig. 2.6(h), it is discarded, and it appears that $\lambda_p = 1\cdot375$.

A static check should be performed on the mechanism in Fig. 2.6(h) to ensure that the yield condition is satisfied. The unknown moments in Fig. 2.6(h) are M_C, M_{DG}, and M_{DE}. Virtual work equations, using displacements corresponding to the elementary mechanisms, may conveniently be used to derive these moments.

Let positive moments be those which cause tension on the upper faces of beams (that is, hogging moments) and on the left faces of columns, as shown by the dotted lines in Fig. 2.6(b). A corresponding sign convention is adopted for hinge rotations. Then, applying the mechanism of Fig. 2.6(d) in a virtual work equation for the loading and moment conditions in Fig. 2.6(h) (that is, with $\lambda = 1·375$) we have

$$6 \times 1·375(10\theta) = -20(\theta) + M_C(-2\theta) + 50(\theta)$$

whence $M_C = -26·25$, which satisfies the yield condition ($M_p = 50$). Similarly, the moment at D in column DE may be calculated by using the sway mechanism in Fig. 2.6(c) in a virtual work equation, giving

$$4 \times 1·375(15\theta) = 20(\theta) - 20(-\theta) + M_{DE}(-\theta) + 20(\theta) + 20(\theta) - 20(-\theta)$$

whence $M_{DE} = 17·5$. The equilibrium condition for the joint D then gives $M_{DG} = 67·5$, which could alternatively be derived by using the elementary mechanism in Fig. 2.6(e). Hence the static check is correct, and $1·375$ is confirmed as the collapse load factor.

An alternative method of carrying out the static check is described in Section 2.5.

Number of elementary mechanisms

It is of help in applying the method of combined mechanisms to be assured that the necessary elementary mechanisms have been considered. If r is the degree of redundancy, p the number of sections at which plastic hinges may occur, and m the number of elementary mechanisms, then

$$m = p - r \tag{2.19}$$

Thus, for the frame in Fig. 2.6(a), $p = 10$, $r = 6$, whence $m = 4$. The proof of Eqn (2.19) is as follows.

Each independent elementary mechanism (including joint mechanisms) represents, by its use in a virtual work equation, an equilibrium condition relating some of the bending moments at the p sections to the external loads for any equilibrium state of the structure. If all the m mechanisms were so used, there would remain a requirement for r additional equations to determine the p bending moments, since the degree of redundancy is known to be r. Hence $m + r = p$.

Combined mechanism method for larger frames

The combined mechanism method is suitable for the analysis of extensive rigid frames, and the frame in Fig. 2.7(a) will be taken as an example, The dimensions, loads, and plastic moments are as shown, and it is required to determine the load factor at which collapse will occur.

There are 22 sections at which it is necessary to allow for the possible occurrence of plastic hinges, and these are indicated in Fig. 2.7(a). The outer columns are pinned at the base, so that the degree of redundancy is 10 (four structural rings giving 12 degrees of redundancy, less two for internal moments eliminated at M and P). With $p = 22$ and $r = 10$, $m = 22 - 10 = 12$. Since there are four joints with more than two members meeting (at C, E, I, and K),

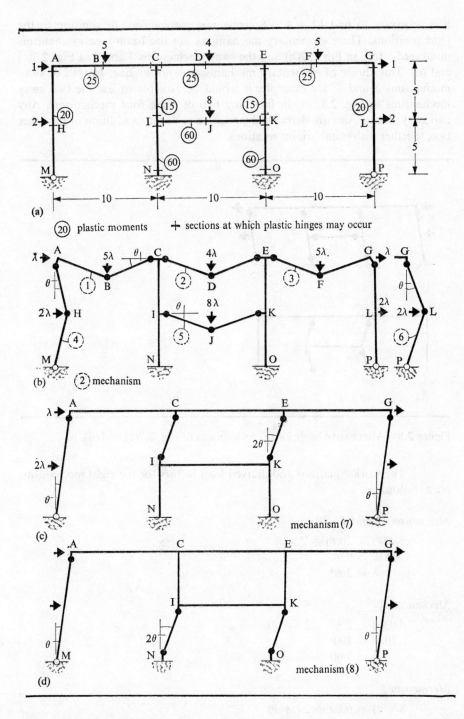

Figure 2.7 (a) Frame dimensions, plastic moments, and loads; (b), (c), (d) Elementary mechanisms

it is necessary to find $12 - 4 = 8$ elementary mechanisms in addition to the joint rotations. These elementary mechanisms are the beam-type mechanisms numbered 1 to 6 in Fig. 2.7(b) and the sway mechanisms 7 and 8 in Figs 2.7(c) and (d). This choice of elementary mechanisms is not unique. Instead of sway mechanisms 7 and 8, for example, it would be possible to use the two sway mechanisms in Fig. 2.8, or, in fact, any two of these four mechanisms. Any two sway mechanisms are derivable by positive or negative addition of the other two, together with suitable joint rotations.

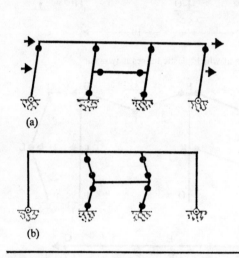

(a)

(b)

Figure 2.8   Alternative basic sway mechanisms to Fig. 2.7(c) and (d).

The work equations and derived load factors for the eight mechanisms are as follows.

Mechanisms 1 and 3

$$5\lambda(5\theta) = 20(\theta) + 25(2\theta + \theta)$$
$$25\lambda\theta = 95\theta$$
$$\lambda = 3{\cdot}80$$

Mechanism 2

$$4\lambda(5\theta) = 25(\theta + 2\theta + \theta)$$
$$20\lambda\theta = 100\theta$$
$$\lambda = 5{\cdot}00$$

Mechanism 5

$$8\lambda(5\theta) = 60(\theta + 2\theta + \theta)$$
$$40\lambda\theta = 240\theta$$
$$\lambda = 6{\cdot}00$$

Mechanisms 4 and 6

$$2\lambda(5\theta) = 20(\theta + 2\theta)$$
$$10\lambda\theta = 60\theta$$
$$\lambda = 6 \cdot 00$$

Mechanism 7

$$2\lambda(10\theta) + 2\lambda(5\theta) + 2\lambda(5\theta) = 20(\theta + \theta) + 15(2\theta + 2\theta + 2\theta + 2\theta)$$
$$40\lambda\theta = 160\theta$$
$$\lambda = 4 \cdot 00$$

Mechanism 8

$$2\lambda(10\theta) + 2\lambda(5\theta) + 2\lambda(5\theta) = 20(\theta + \theta) + 60(2\theta + 2\theta + 2\theta + 2\theta)$$
$$40\lambda\theta = 520\theta$$
$$\lambda = 13 \cdot 00$$

It appears advantageous to combine mechanisms 1, 3, and 7, since these all have load factors at or near to the lowest value of 3·80. A joint rotation will also be involved at E, and since in mechanism 7 column EK rotates through 2θ, the elementary mechanism involving beam EG (mechanism 3) must be included with rotations 2θ instead of θ to allow the cancellation of plastic hinges. It is by the cancellation of plastic hinge work when combining mechanisms that the total internal work is reduced, and the load factor λ thereby decreased. Denoting the combined mechanism shown in Fig. 2.9(a) by C1, we have C1 = (1) + 2(3) + (7), whence

$$(25 + 2 \times 25 + 40)(\lambda\theta) = (95 + 2 \times 95 + 160)\theta - 2 \times 20(\theta)$$
$$- 25(2\theta) - 15(2\theta) + 25(2\theta)$$
$$115\lambda\theta = 375\theta$$
$$\lambda = 3 \cdot 26$$

In deriving the work equation, the terms on the left-hand side are obtained simply by adding the work equations for the elementary mechanisms. The three negative terms on the right-hand side are respectively for the cancellation of plastic hinges at A, at E in beam EG, and at E in column EK. The final positive term is for the plastic hinge rotation at E in beam CE.

It is now necessary to explore whether further combinations should be made with beam-type mechanisms 2, 4, 5, and 6. Since the column GP has lost its restraining plastic hinge at G, and since this hinge now operates in the opposite sense, it is intuitively obvious that mechanism 6 ought probably to be included, giving the new combined mechanism C2 in Fig. 2.9(b), where C2 = C1 + 3(6). Mechanism 6 is multiplied by three to cancel the hinge at G where there is a rotation of 3θ. Hence

$$(115 + 3 \times 10)\lambda\theta = (375 + 3 \times 60)\theta - 2 \times 20(3\theta)$$
$$145\lambda\theta = 435\theta$$
$$\lambda = 3 \cdot 00$$

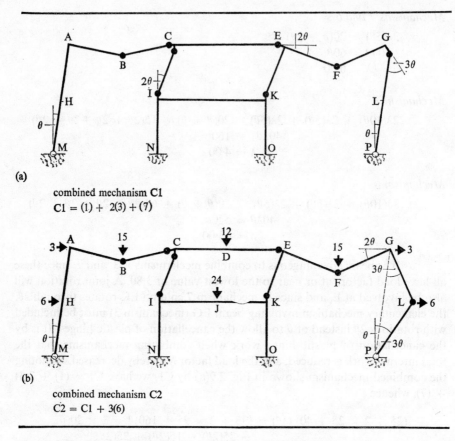

combined mechanism **C1**

C1 = (1) + 2(3) + (7)

combined mechanism **C2**

C2 = C1 + 3(6)

Figure 2.9   Combined mechanisms for frame in Fig. 2.7(a)

The effect of adding other beam-type mechanisms, or sway mechanism 8 [Fig. 2.7(d)], is to increase the load factor, whence the plastic collapse load factor appears to be 3·00. To attempt all possible combinations of mechanisms is excessively tedious, and at some stage it is usually profitable to make a static check of the bending moments throughout the structure rather than to continue testing further combinations. A general procedure is described in Section 2.5 using the present frame as an example, and it is there confirmed that the plastic collapse load factor is 3·00 for this frame.

The number of plastic hinges in the mechanism in Fig. 2.9(b) is 8, compared with a degree of redundancy of 10, so that it is to be expected that there will be three unknown internal reactions when deriving the bending moments. In fact, the part of the structure NIKO contains no plastic hinge, so that the bending moments in IK, IN, and KO will remain statically undetermined. It is necessary when carrying out a static check to confirm that it is possible to postulate a bending moment distribution satisfying the yield condition, even in a statically undetermined part of a structure, and the method described in Section 2.5 allows this to be done.

2.5   Bending moments in collapse mechanisms

It has already been shown (Section 2.4) that the bending moments, other than the plastic hinge values, in a plastic mechanism may be obtained from virtual work equations using the elementary mechanisms. An alternative procedure is to consider, step by step, the equilibrium of each part of the structure, and this method is now described.

Suppose that the beam AC in Fig. 2.10(a), with central load applied at B, is contained in a collapse mechanism, and that the bending moments at B and C are known to be as indicated. This shows that there is an anticlockwise shear force in BC of (50 + 30)/5 = 16, whence the clockwise shear force in AB must be 9 in order to support the applied load of 25 units. The total anti-clockwise moment acting on the two ends of AB must be 9 × 5 = 45, and, since an anticlockwise moment of 50 acts at E, there must be a clockwise moment of 5 acting on the beam at A as shown.

A different case is shown in Fig. 2.10(b), where the two end moments are known. If the applied load of 20 were absent, there would be an anti-clockwise shear force of (40 − 10)/10 = 3. The addition of the central load of 20 units adds a clockwise shear force in AB of 10 and an anticlockwise shear force in BC, also of 10, so that the net shear forces in the two halves of the beam are as shown. The central bending moment can then be calculated by considering either the shear force in AB or that in BC.

This sort of procedure is readily applied to the structure in Fig. 2.7(a) failing according to the mechanism in Fig. 2.9(b), and the complete solution is

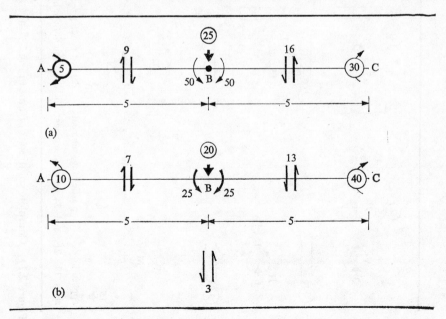

(a)

(b)

Figure 2.10   Use of shear forces in calculating equilibrium moments

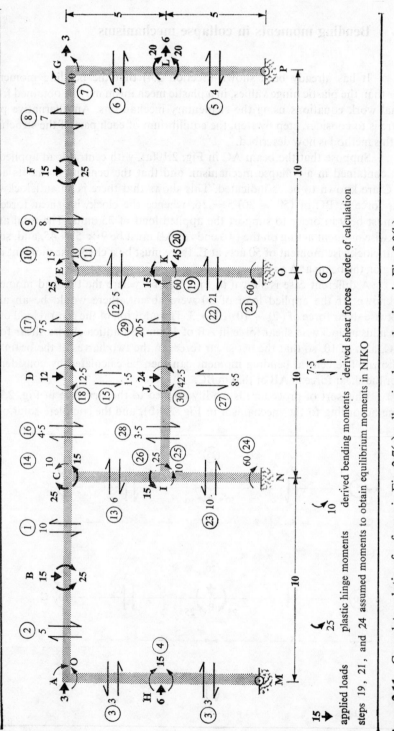

Figure 2.11 Complete solution for frame in Fig. 2.7(a) collapsing by mechanism in Fig. 2.9(b).

given in Fig. 2.11. The initial information is shown bold, and one order in which the calculations may be made is indicated by the circled numbers. Some variation in the order of working is possible. It is suggested that the reader draw out the initial information in the manner shown, and himself add the various steps to obtain a better understanding of the method.

The shear forces in the beam AC are first derived, enabling the bending moment at A to be established. Since this moment is found to be zero, the horizontal load of 6 units applied at H is equally divided between the upper and lower lengths of column AM, thus enabling the moment at H to be obtained.

The working is now continued along members PG, GE, and EK in that order as far as step 12. Since the total horizontal load applied above the upper-storey level is 6 units and considering the shear forces in the other columns, the clockwise shear force in column CI is $6 + 3 - 5 + 2 = 6$. All moments and shear forces in beam CE may now be derived, finishing at step 18.

This completes the statically determined internal forces. To postulate a set of moments satisfying yield and equilibrium in the members NIKO, plastic hinge moments are introduced at one or two sections where, if collapse of this part of the structure *were* to occur, one might expect plastic hinges to arise. Plastic hinge moments of value 60 are therefore introduced first at K in beam IK (step 19) and then at O in column KO (step 21). The high anticlockwise shear force then arising in column IN indicates an appreciable clockwise moment at either I or N, and so arbitrarily a clockwise moment of 60 (equal to the plastic value) is entered at N (step 24). The moments may then be completed throughout the structure, and since they satisfy the yield criterion, the collapse load factor $\lambda_p = 3 \cdot 00$ is confirmed.

It is not always possible to obtain moments satisfying the yield criterion at a first attempt, but if the collapse mode is correct, a little trial and error soon leads to a solution. If a solution is not obtained quickly in any part of the structure that is still 'rigid', the effect of modifying the mechanism by combining with a suitable elementary mechanism should be tried.

2.6   Distributed loads

In methods based on the upper bound theorem, it is most convenient in the first instance to assume that sagging beam hinges under distributed loads can occur only at the end or at mid-span. A uniformly distributed load of w per unit length applied to a beam of length L [Fig. 2.12(a)] may then be replaced by a central concentrated load $wL/2$ and two end loads of $wL/4$, as shown in Fig. 2.12(b). The external work for an elementary collapse mechanism in which both halves of the beam rotate through θ is then $(wL^2/4)\theta$ in each case. When the collapse load has been obtained for the structure for this modified loading, revised positions are calculated for the sagging plastic hinges, and a modified load factor is calculated. This process may be illustrated by reference to the two-bay frame in Fig. 2.13(a).

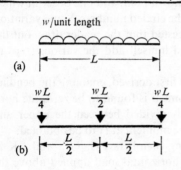

Figure 2.12 (a) Uniform loading of beam; (b) equivalent point loading

When the uniformly distributed loads in Fig. 2.13(a) given as loads per unit length are replaced by equivalent concentrated loads, the loading and the frame are identical with the example previously considered [Fig. 2.6(a)], apart from the values of the axial loads in the columns, which have no effect on the failure load. Hence the collapse mechanism is as shown in Fig. 2.13(b) and $\lambda = 1 \cdot 375$. Since the plastic hinge moment at F in beam DG is not necessarily the maximum sagging moment, this load factor is only an upper bound on the true load factor at failure.

Considering the beam DG in further detail [Fig. 2.13(c)], the vertical reaction R_G acting on the beam at G is given by taking moments about the mid-span F, whence

$$15R_G = 80 + 20 + (1 \cdot 100 \times 15) \times 7 \cdot 5$$

or $R_G = 14 \cdot 917.$

If F′, distance x from G, is the position of zero shear force (and therefore maximum bending moment) in the beam, $1 \cdot 100x = 14 \cdot 917$, whence $x = 13 \cdot 561$. The maximum sagging moment is therefore of magnitude $14 \cdot 917x - 1 \cdot 100x^2/2 - 20 = 81 \cdot 14$, compared with a plastic moment of 80. A lower bound to the failure load factor is therefore $1 \cdot 375 \times 80/81 \cdot 14 = 1 \cdot 356$, so that $1 \cdot 356 \leqslant \lambda_p \leqslant 1 \cdot 375$.

The mechanism of Fig. 2.13(b) is now revised so that the sagging hinge in the beam DG occurs at F′, the position obtained for the maximum sagging moment. This gives the mechanism in Fig. 2.13(d) in which the plastic hinge rotations at F′ and G are each $\theta + [(30 - x)/x]\theta = (30/x)\theta$. When $x = 13 \cdot 561$, the work equation gives $\lambda = 1 \cdot 365$, which is therefore the revised estimate of the failure load.

Since the load factor used in calculating the position of F′ was in fact incorrect ($1 \cdot 375$ compared with $1 \cdot 365$), the true value of x differs slightly from that assumed. This process of estimating failure loads when uniformly distributed loads are present is therefore, strictly speaking, one of successive

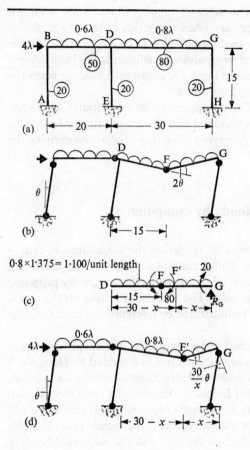

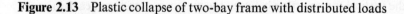

Figure 2.13 Plastic collapse of two-bay frame with distributed loads

approximation, but it is so rapidly convergent that only one cycle of calculation, as just performed, is usually required.

An alternative method is to derive λ in terms of x from the work equation, and then to minimize λ with respect to x. The work equation for the mechanism in Fig. 2.13(d) is

$$4\lambda(15\theta) + 24\lambda\left(\frac{30-x}{2}\theta\right) = 20(\theta + \theta + \theta + \theta) + 50(\theta) + 80\left(\frac{30}{x}\theta\right)$$
$$+ 20\left(\frac{30}{x}\theta\right)$$

giving $\lambda = \dfrac{5}{6}\left\{\dfrac{300 + 13x}{(35 - x)x}\right\}$ (2.20)

Putting $d\lambda/dx = 0$ gives $x = 13 \cdot 532$, which differs negligibly from the first estimate obtained of $13 \cdot 561$, and leads to a negligibly different failure load. (To six-figure accuracy, the correct λ is $1 \cdot 36514$ compared with $1 \cdot 36520$ obtained for $x = 13 \cdot 561$). When there are a number of sagging hinges with unknown positions $x_1, x_2, \ldots$, it becomes necessary when using this method to minimize with respect to all these variables by putting $\partial \lambda / \partial x_1 = 0$, $\partial \lambda / \partial x_2 = 0$, etc. Hence the first method is generally to be preferred, although a formula has been derived (Horne[3]) for exactly determining failure loads for distributed loading when the failure loads under equivalent loading are known.

It is possible for a sagging hinge to appear under a distributed load when there is no sagging hinge under the equivalent mid-span concentrated load. This emphasizes the importance of finally carrying out a static check and deriving the bending moments throughout the structure.

2.7 Calculation of failure loads by computer

The most straightforward way of rendering the calculation of failure loads automatic for calculation by digital computer is to use the maximum theorem. The general basis of the method is similar to that originally proposed as a hand method by Neal and Symonds.[1] The method of setting up the necessary equations and inequalities will be illustrated by reference to the rectangular grid shown in plan in Fig. 2.14.

The grid members are assumed to be continuous and simply supported at their ends. The general plastic theory of grids was first treated by Heyman[4-6] and involves the plastic failure of members under combined bending and torsion. If the members are of standard rolled I-section, the torsional resistance is small compared with resistance to bending and may be neglected. This procedure will be adopted here. The only important internal forces in the members are then the shear forces perpendicular to the plane of the grid and the associated bending moments. The number of redundancies is equal to the number of grid joints, since, if the vertical forces acting between the members crossing at each joint are specified, the end support reactions and all bending moments may be calculated. For the grid in Fig. 2.14(a) there are therefore 12 degrees of redundancy.

Let the plastic moments of the members be M_{pA}, M_{pB}, M_{pC}, and M_{pD} as shown and let vertical loads $\lambda W_1, \lambda W_2, \ldots, \lambda W_{12}$ be applied at the nodes only. Let M_{1x} and M_{1y} be the sagging bending moments at joint 1 in members RS and TU respectively, with a corresponding notation at the other joints. Since there are 12 degrees of redundancy and 24 unknown moments (two at each joint), there must be 12 independent mechanisms, each representing an equation of equilibrium. Taking these 12 mechanisms as those obtained by the descent *by itself* of each joint, the mechanism for joint 1 has hinges in member RS at 1 and 2 and in member TU at 1 and 5. Allowing for the signs of the virtual rotations, the virtual work equation for a descent Δ of joint 1 becomes:

$$\lambda W_1 \Delta = M_{1x} \left(\frac{2\Delta}{a} \right) - M_{2x} \left(\frac{\Delta}{a} \right) + M_{1y} \left(\frac{2\Delta}{b} \right) - M_{5y} \left(\frac{\Delta}{b} \right)$$

or $\qquad \lambda W_1 ab = 2b M_{1x} - b M_{2x} + 2a M_{1y} - a M_{5y}$ \hfill (2.21)

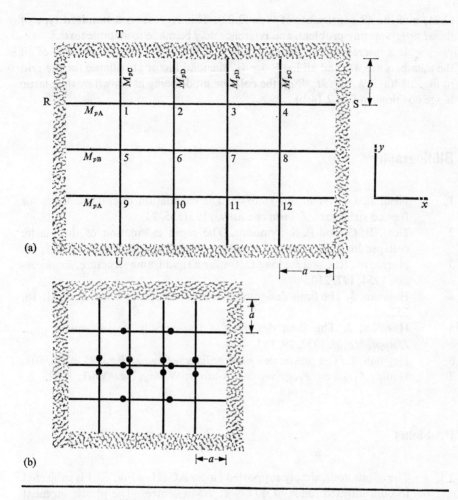

Figure 2.14 (a) Simply supported grid; (b) Failure mechanism with equal spacing, equal plastic moments, and equal loads at all joints

The virtual work equations for the remaining 11 joints complete the required 12 equations of equilibrium.

The satisfaction of the yield conditions involves 24 statements of the form

$$|M_{1x}| \leqslant M_{pA}$$

For computational purposes, it is necessary to replace the inequalities referred to the modulus of M_{1x} by the two inequalities

$$\left. \begin{array}{l} M_{1x} \leqslant M_{pA} \\ -M_{1x} \leqslant M_{pA} \end{array} \right\} \tag{2.22}$$

leading to a total of 48 inequalities. The load factor at which the grid will just collapse is then the maximum value of λ, subject to the 12 constraints of type

5—P.T.S.

(2.21) and the 48 constraints of type (2.22). This represents a standard type of linear programming problem, and reference may be made to suitable texts.[7]

It is interesting to note that, if $a = b$, and if the plastic moments of all the members are M_p and all loads are W, the load factor at collapse for the grid in Fig. 2.14(a) is $\lambda = \frac{5}{6}(M_p/Wa)$, the collapse mode being as shown by the plastic hinge positions in Fig. 2.14(b).

Bibliography

1   Neal, B. G. and P. S. Symonds. The calculation of collapse loads for framed structures. *J. Instn civ. Engrs*, 1951, **35**, 21.
2   Neal, B. G. and P. S. Symonds. The rapid calculation of the plastic collapse load for a framed structure. *Proc. Instn civ. Engrs*, 1952, **1**, 58.
3   Horne, M. R. Collapse load factor for a rigid frame structure. *Engineering*, 1954, **177**, 210.
4   Heyman, J. The limit design of space frames. *J. Appl. Mech.*, 1951, **18**, 157.
5   Heyman, J. The limit design of a transversely loaded square grid. *J. Appl. Mech.*, 1952, **19**, 153.
6   Heyman, J., The plastic design of grillages. *Engineering*, 1953, **176**, 804.
7   Hadley, G. *Linear Programming*. Addison-Wesley, New York, 1962.

Problems

2.1   The continuous, simply supported girder AGHJ in Fig. E2.1 is subjected to concentrated loads of 40 kN at 2·5 m centres. The plastic moment is uniform at 700 kNm over the central span, but varies over the end spans as follows:

Section	A	B	C	D	E	F	G
M_p (kNm)	250	300	350	400	500	600	700

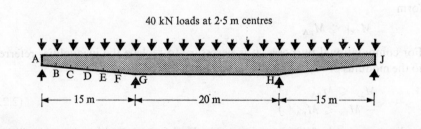

40 kN loads at 2·5 m centres

Figure E2.1

Draw the free moment diagram, and, by assuming a sagging plastic hinge at D, show that the load factor at collapse lies in the range 1·345–1·667. Show that the correct value is 1·460.

2.2 The rigid-based frame in Fig. E2.2 is initially designed with a uniform section so that it just collapses under the loads shown. Determine the value of the plastic moment.

The frame is now redesigned with a uniform basic section, but haunches are provided from points 1 m below the eaves at B and D to the first loading points B_1 and D_1 in the rafters. Determine the plastic moment of the basic section and the maximum moment in the haunch.

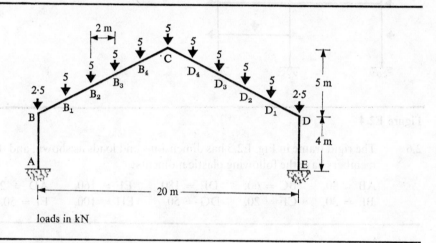

loads in kN

Figure E2.2

2.3 Repeat the designs in Example 2.2, assuming that the frame is pinned at A and E.

2.4 The rigid-based frame in Fig. E2.3 has a uniform plastic moment M_p.
(a) Calculate the value of w at collapse in terms of M_p, on the assumption that any sagging hinges within the lengths of beams AC and CE occur at mid-span.

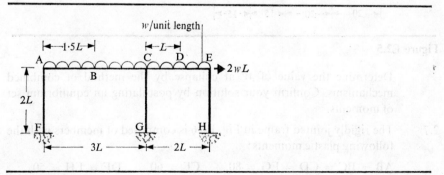

Figure E2.3

(b)   Use the solution of (a) to refine the collapse mechanism and recalculate the value of *w*.

2.5   The frame in Fig. E2.4 is to be analysed by the method of combined mechanisms. State the degree of redundancy *p* and the number of independent mechanisms *m*. Sketch a set of independent mechanisms.

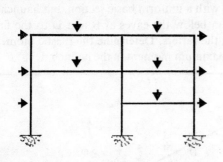

Figure E2.4

2.6   The rigid frame in Fig. E2.5 has dimensions and loads as shown, and the members have the following plastic moments:

AB = 80,   BC = 60,   DE = 180,   EF = 160,   AD = 20,
BE = 30,   CF = 20,   DG = 50,   EH = 100,   FI = 50.

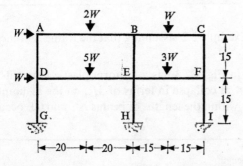

Figure E2.5

Determine the value of *W* at collapse by the method of combined mechanisms. Confirm your solution by postulating an equilibrium set of moments.

2.7   The rigidly jointed frame in Fig. E2.6 is composed of members with the following plastic moments:

AB = BC = CD = EG = 80,   CE = 60,   DF = FH = 50,
EF = 150.

Determine the factor by which the loads shown would have to be multiplied for collapse just to occur. Confirm your solution by postulating an equilibrium set of moments.

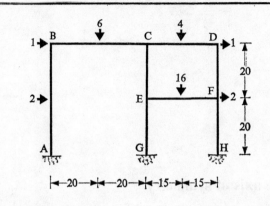

Figure E2.6

Chapter 3
Plastic moments under shear force and axial load

3.1 Sections with one axis of symmetry

The T-section in Fig. 3.1(a) has a plane of symmetry through the principal axis YY, while the other principal axis XX, passing through the centroid G, is not an axis of symmetry. We consider bending *in* the plane of symmetry, *about* the axis XX. When this section is under a sagging bending moment which causes elastic stresses only, the stress distribution is as shown, the neutral axis coinciding with XX. The yield moment M_y is that which causes yield in tension at the bottom of the web, that is $\sigma_2 = \sigma_y$. The value of the elastic modulus Z_e for the section with the dimensions shown is found to be $10^3 \times 86 \cdot 8$ mm³, so that $M_y = 10^3 \times 86 \cdot 8 \sigma_y$.

When the bending moment is increased beyond M_y a plastic zone forms in tension at the bottom of the web [Fig. 3.1(b)]. The neutral axis AA now becomes distinct from the centroidal axis XX. At still larger moments a plastic zone forms in compression in the flange, until finally the plastic moment M_p is reached [Fig. 3.1(c)]. The position of the neutral axis AA [defined by the dimension y_p in Fig. 3.1(c)] must be such that there is no resultant axial force in the member, that is, it is the *equal area axis*, dividing the section into two equal areas. Hence

$$160 y_p = 160(20 - y_p) + 10 \times 160$$
$$y_p = 15 \text{ mm}$$

The plastic modulus Z_p is obtained by taking the first moment of area about the equal area axis, regarding the moments on either side as positive, so that

$$Z_p = (160 \times 15 \times 7 \cdot 5) + (160 \times 5 \times 2 \cdot 5) + (10 \times 160 \times 85)$$
$$Z_p = 10^3 \times 156 \text{ mm}^3$$
or $\qquad M_p = 10^3 \times 156 \sigma_y$

It may be noted that taking moments about any axis (including axis XX) leads to an identical result, since the forces on the cross section reduce to a pure moment in the plane of symmetry.

The shape factor for this section is $\nu = Z_p/Z_e = 1\cdot80$.

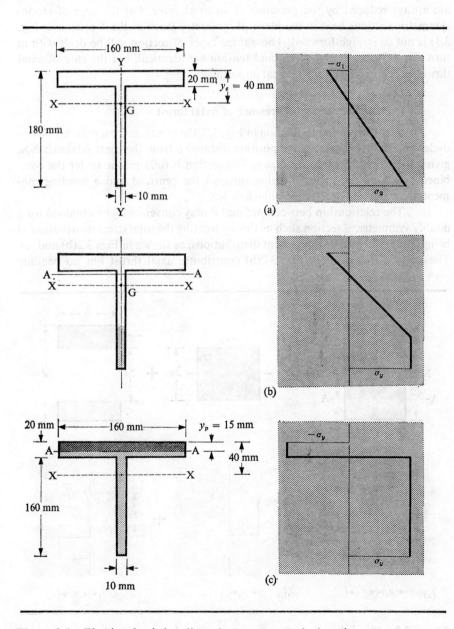

Figure 3.1 Elastic–plastic bending of monosymmetrical section

3.2   Effect of axial load on plastic moment

In the presence of axial load (either tensile or compressive), the neutral axis no longer divides a section into two equal areas. The plastic moments of doubly symmetrical sections (such as rectangular sections and rolled I-sections) are always reduced by the presence of an axial force, but the case of mono-symmetric sections bent in the plane of symmetry (such as the T-section in Fig. 3.1) is not so straightforward. These three types of section will be dealt with in turn. The effects of axial thrust and tension are identical, but the case of axial thrust is usually of greater practical importance.

Rectangular section in presence of axial thrust

For the rectangular section in Fig. 3.2, the plastic neutral axis AA moves under an axial thrust P to some position distance a from the centroidal axis XX, giving the stress distribution shown. The section is fully plastic under the combined action of the thrust P acting through the centroid and a bending moment M_p' acting about the centroidal axis XX.

The relationship between M_p' and P may conveniently be obtained for a doubly symmetrical section such as this by treating the total stress distribution as being composed of two component distributions, as shown in Figs 3.2(b) and (c). The stress distribution in Fig. 3.2(b) contributes axial thrust but no bending

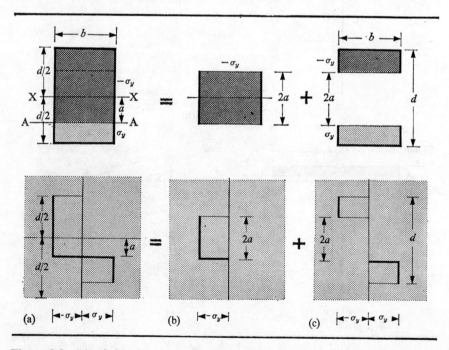

Figure 3.2   Plastic bending of rectangular section in presence of axial thrust

moment, while that in Fig. 3.2(c) contributes bending moment but no thrust. Hence

$$P = 2ab\sigma_y \tag{3.1}$$

$$M_p' = \frac{bd^2}{4} \sigma_y - ba^2\sigma_y \tag{3.2}$$

where M_p' is obtained by deducting from the plastic moment for the complete section the 'lost' moment $ba^2\sigma_y$ due to the removal of a depth $2a$.

If the section were plastic under an axial thrust only, the value of the thrust would be $P_p = A\sigma_y = bd\sigma_y$, where A is the area of cross section. The thrust P_p is known as the *squash load*, since, if a short specimen were subjected to such a load, it would deform axially in pure plastic deformation. (A specimen of appreciable length would not deform uniformly but would buckle.) The ratio P/P_p [$= (P/A)/\sigma_y$ = mean axial stress/yield stress] will be denoted by n (the *squash load ratio*) so that, from Eqn (3.1), $a = nd/2$, and putting $M_p' = Z_p'\sigma_y$, where Z_p' is a modified plastic modulus allowing for the effect of axial thrust,

$$Z_p' = (1 - n^2)Z_p \tag{3.3}$$

Hence the plastic moment $M_p'[=(1 - n^2)Z_p\sigma_y]$ varies parabolically with the thrust $P = An\sigma_y$, as shown by the upper curve in Fig. 3.3, becoming zero at the squash load. This behaviour may be compared with the linear reduction of the moment M_y' at first yield, as given by the lower straight line in Fig. 3.3 $[M_y' = (1 - n)Z_e\sigma_y = \frac{2}{3}(1 - n)Z_p\sigma_y)]$.

I-section in presence of axial thrust

A symmetrical I-section may be treated similarly to a rectangular section. When the plastic neutral axis remains in the web (Fig. 3.4), we have

$$P = 2at_w\sigma_y \tag{3.4}$$

$$M_p' = M_p - t_wa^2\sigma_y \tag{3.5}$$

where M_p is the plastic moment under zero axial load. If the total area of the section is A and the squash load ratio is again denoted by n, so that $P = An\sigma_y$, it follows from Eqn (3.4) that $a = (A/2t_w)n$, whence from Eqn (3.5), if $M_p' = Z_p'\sigma_y$,

$$Z_p' = Z_p - \left(\frac{A^2}{4t_w}\right) n^2 \tag{3.6}$$

This result remains valid until the plastic neutral axis reaches the flange, that is until $a = d/2 - t_f$ or $n = t_w(d - 2t_f)/A$. The section is then conveniently divided as in Fig. 3.5; from Fig. 3.5(b),

$$P = \{A - (d - 2a)b\}\sigma_y \tag{3.7}$$

and from Fig. 3.5(c),

$$M_p' = 2[(d/2 - a)b \{\tfrac{1}{2}(d/2 + a)\}]\sigma_y \tag{3.8}$$

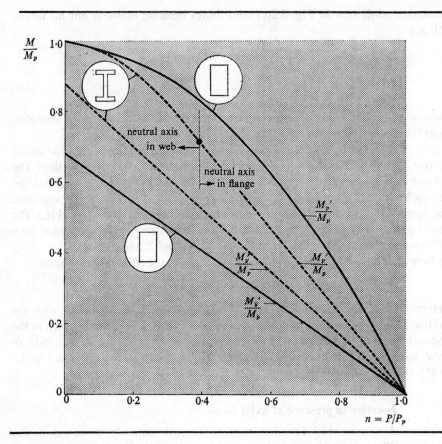

Figure 3.3 Effect of axial force on plastic moments in rectangular and I-sections

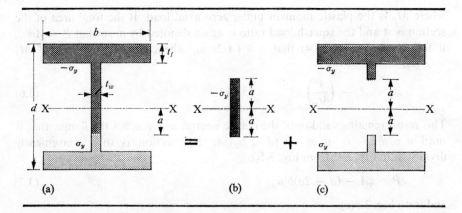

Figure 3.4 I-section under axial thrust and bending–neutral axis in web

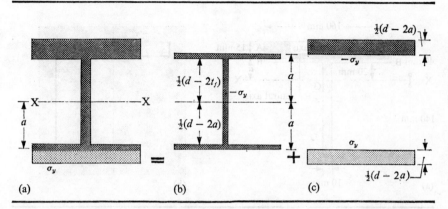

Figure 3.5 I-section under axial thrust and bending–neutral axis in flange

Substituting $P = An\sigma_y$ in Eqn (3.7) gives $1 - n = (d - 2a)b/A$, whence, from Eqn (3.8)

$$Z'_p = \frac{A^2}{4b}(1 - n)\left(\frac{2bd}{A} - 1 + n\right) \tag{3.9}$$

Formulae of the type of Eqns (3.6) and (3.9) are available for standard sections in the handbooks. The variation of plastic moment $M'_p = Z'_p\sigma_y$ with axial thrust $P = nP_p$ for a typical I-section (Fig. 3.3) lies below the curve for a rectangular section, but even for an I-section the reduction in plastic moment is very small for small values of the thrust. For a mean axial stress equal to 10% of the yield stress (that is, for $n = 0\cdot1$), the plastic moment of a rectangular section is reduced by only 1% and that of a typical I-section by about 2%. It will therefore be found justifiable in many problems to ignore the effect of axial forces on plastic moments, although it must always be remembered that for slender members, axial thrust may cause serious instability effects (Chapter 6).

Unsymmetrical sections in presence of axial thrust

At zero axial load, the neutral axis at full plasticity for the I-section in Fig. 3.1 is the equal area axis AA in Fig. 3.1(c). Under axial thrust or tension the neutral axis may lie in the web or in the flange, depending on the sense and value of the axial load and the sign (hogging or sagging) of the bending moment. Suppose the section is under sagging bending moment and that the neutral axis BB [Fig. 3.6(a)] is in the flange at a distance a from the centroidal axis. The stress distribution is as shown in Fig. 3.6(b). If $n\sigma_y$ is the mean axial compressive stress, since the area of cross section is 4800 mm^2, it follows that

$$4800n\sigma_y = 160(40 - a)\sigma_y - 160(a - 20)\sigma_y - 10 \times 160\sigma_y.$$

Hence $a = 25 - 15n$. Since $20 \leqslant a \leqslant 40$ for the neutral axis to be in the flange, $-1 \leqslant n \leqslant \frac{1}{3}$.

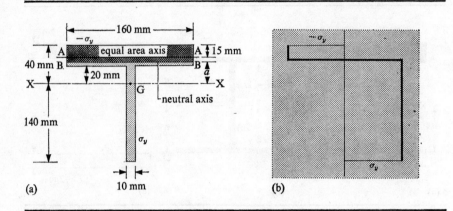

Figure 3.6   Bending of monosymmetrical section in presence of axial force

If the sagging bending moment M'_p is considered to act about the centroidal axis XX, then

$$M'_p = \frac{160(40^2 - a^2)}{2}\sigma_y - \frac{160(a^2 - 20^2)}{2}\sigma_y + \frac{10(140^2 - 20^2)}{2}\sigma_y$$

The plastic moment M_p under zero axial load is $10^3 \times 156\sigma_y$, so that after simplification,

$$\frac{M'_p}{M_p} = \frac{1600 - a^2}{975} = \tfrac{1}{13}(1 + n)(13 - 3n) \tag{3.10}$$

When $n = -1$, $a = 40$ and $M'_p = 0$, that is, the whole section is plastic in tension. Taking the other limit for the neutral axis lying in the flange ($n = \tfrac{1}{3}$, $a = 20$) gives $M'_p = \tfrac{16}{13}M_p$, so that the application of axial load causes an increase in the plastic moment. If at first this appears anomalous, it should be recollected that in the elastic range, the load at first yield in an unsymmetrical section may also be increased by axial force.

If the bending moment is measured in the present case about the equal area axis AA instead of the centroidal axis XX, it is found that

$$\frac{M'_p}{M_p} = 1 - \tfrac{3}{13}n^2 \tag{3.11}$$

provided the neutral axis remains in the flange, that is, $-1 \leqslant n \leqslant \tfrac{1}{3}$. Hence M'_p is in this case a maximum when $n = 0$. On the other hand, when $n = -1$, that is, when the whole section is plastic in tension, the plastic moment is no longer zero, but is of value $\tfrac{10}{13}M_p$. It does not matter greatly about which axis the bending moment is in fact measured, provided consistency of treatment is maintained.

The solution for the case when the neutral axis is in the web ($\tfrac{1}{3} \leqslant n \leqslant 1$) is similar and gives the result

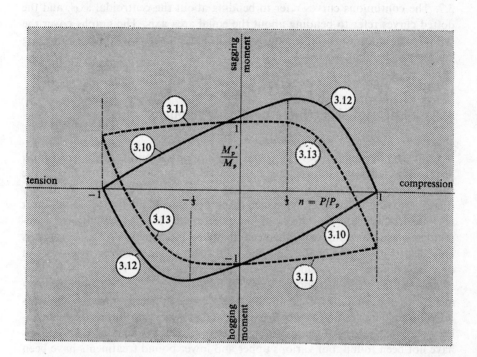

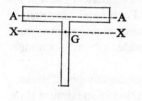

——————— moments taken about centroidal axis XX.

-------- moments taken about equal area axis AA

(3.11) equation numbers

Figure 3.7 Variation of plastic moment with axial force for section in Fig. 3.6(a)

$$\frac{M_p'}{M_p} = \tfrac{8}{13}(1-n)(1+6n) \tag{3.12}$$

about the centroidal axis and

$$\frac{M_p'}{M_p} = \frac{8 - 5n - 6n^2}{13} \tag{3.13}$$

about the equal area axis. Complete results are represented graphically in Fig.
3.7. The continuous curves refer to bending about the centroidal axis, and the
dotted curves refer to bending about the equal area axis. The graphs cover the
full possible range for axial tension as well as axial thrust, and for both hogging
and sagging bending moments.

3.3   Effect of shear force on plastic moment

The Bernoulli theory of elastic bending of a beam is only correct for a
beam subjected to a uniform moment. Unless the moments are applied at the
ends in the same way as the elastic stress distributions derived by the theory
(that is, varying linearly with distance from the centroidal axis), the beam must
be long to eliminate the effects of the precise end loading conditions. Non-
uniform elastic bending, with the accompanying introduction of shear force, is
only dealt with approximately by Bernoulli theory, and a correct solution can
only be obtained by considering each problem in detail. There is no unique
solution to the stress distribution across a *section* of an elastic member in terms
merely of the bending moment and shear force at that section—the stress
distribution depends on the whole loading and support conditions for the member. The same is true for non-uniform elastic–plastic bending. Exact solutions
have not been found, but various upper and lower bound treatments have been
suggested (see, for example, Green[1] and Horne[2]), and a numerical method for
obtaining the exact solution has been described (Neal[3]). Fortunately, in practical
problems the effect of shear is usually small, and solutions giving approximate
lower bounds are readily obtained. Reference may be made to Neal[4] for a more
extensive discussion.

It is necessary to adopt some criterion for yield under combined stresses.
The two most commonly used are those of Tresca (criterion of maximum shear
stress) and von Mises (strain energy of distortion, or octahedral shear stress). As
applied to a plane stress condition in which one plane is acted upon by a normal
stress σ and a shear stress τ, while the plane at right angles is acted upon by the
complementary shear stress $-\tau$ only (normal stress zero), the Tresca yield criterion takes the form

$$\sigma^2 + 4\tau^2 = \sigma_y^2 = 4\tau_y^2 \qquad (3.14)$$

where σ_y is the yield stress in pure tension or compression and τ_y is the yield
stress in pure shear. The von Mises criterion takes the form

$$\sigma^2 + 3\tau^2 = \sigma_y^2 = 3\tau_y^2 \qquad (3.15)$$

Both criteria may be expressed by the relationship

$$\left(\frac{\sigma}{\sigma_y}\right)^2 + \left(\frac{\tau}{\tau_y}\right)^2 = 1 \qquad (3.16)$$

which is therefore used throughout the following treatment.

Effect of shear force in rectangular sections

Let the cantilever AB in Fig. 3.8, with rectangular section $b \times d$, be fixed rigidly at end B and let it support a single vertical load F at end A. Plastic zones in direct tension and compression will spread to the right of section A_1A_1, distance x_1 from end A, where

$$Fx_1 = M_y = \left(\frac{bd^2}{6}\right)\sigma_y \tag{3.17}$$

Using the usual simple theory of elastic bending, the normal stresses on section A_1A_1 will be distributed linearly as shown in Fig. 3.9(a), while the shear stresses will be distributed parabolically [Fig. 3.9(b)]. The greatest shear stress on a

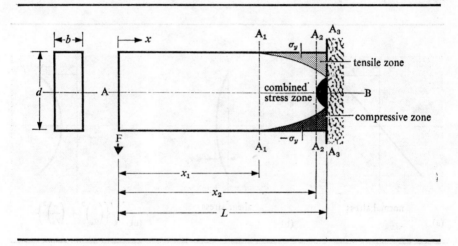

Figure 3.8   Effect of shear stresses on plastic zones in cantilever

vertical plane occurs on the central axis and is of value $\frac{3}{2}(F/bd)$. For this to be less than τ_y, the yield stress in pure shear, it follows from Eqn (3.17) that $x_1/d > \frac{1}{4}(\sigma_y/\tau_y)$, that is, $x_1/d > 0.500$ for the Tresca criterion and $x_1/d > 0.433$ for the von Mises criterion. Hence yield will always occur in the outermost fibres in pure tension and compression before it occurs in the central fibres in pure shear in any situation in which simple bending theory is at all applicable. The condition for elastic behaviour, namely $(\sigma/\sigma_y)^2 + (\tau/\tau_y)^2 < 1$ [see Eqn (3.16)] is found to be satisfied everywhere except in the extreme fibres [Fig. 3.9(c)].

As the tensile and compressive plastic zones grow for $x > x_1$, the shear force is taken entirely by the elastic core, and the shear stress on the central axis increases until it reaches τ_y. This occurs at some section A_2A_2 (Fig. 3.8) the normal and shear stress distributions being as shown in Figs 3.10(a) and (b). Shear stresses are distributed parabolically across the elastic core of depth $2y_o$, whence

$$Fx_2 = M = \frac{b}{4}(d^2 - 4y_o^2)\sigma_y + \frac{2}{3}by_o^2\sigma_y$$

that is, $M = b\left(\dfrac{d^2}{4} - \dfrac{y_o^2}{3}\right)\sigma_y$ (3.18)

$F = \tfrac{4}{3}by_o\,\tau_y$ (3.19)

The value of $\sqrt{\{(\sigma/\sigma_y)^2 + (\tau/\tau_y)^2\}}$ is shown in Fig. 3.10(c), and it is seen that the yield criterion is very nearly reached over the entire cross section. Hence the combinations of M and F defined by Eqns (3.18) and (3.19) are nearly sufficient to produce full plasticity. Putting

$$M_p = \frac{bd^2}{4}\sigma_y$$

and $F_p = bd\tau_y$ (3.20)

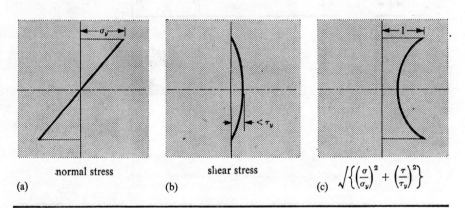

normal stress   shear stress   (c) $\sqrt{\left\{\left(\dfrac{\sigma}{\sigma_y}\right)^2 + \left(\dfrac{\tau}{\tau_y}\right)^2\right\}}$

(a)   (b)

Figure 3.9   Distribution of stresses at section A_1A_1 in Fig. 3.8

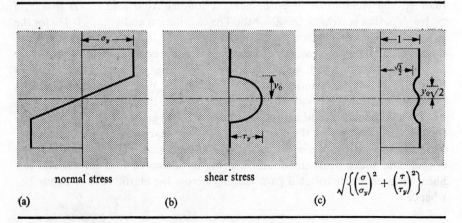

normal stress   shear stress   (c) $\sqrt{\left\{\left(\dfrac{\sigma}{\sigma_y}\right)^2 + \left(\dfrac{\tau}{\tau_y}\right)^2\right\}}$

(a)   (b)

Figure 3.10   Distribution of stresses at section A_2A_2 in Fig. 3.8

the elimination of y_o between Eqns (3.18) and (3.19) leads to the following lower bound estimate for the reduced plastic moment M'_p :

$$\frac{M'_p}{M_p} = 1 - \frac{3}{4}\left(\frac{F}{F_p}\right)^2 \tag{3.21}$$

This applies for values of F up to that at which $y_o = d/2$, that is, from Eqn (3.19) it applies when

$$\frac{F}{F_p} \leqslant \frac{2}{3} \tag{3.22}$$

Equation (3.21) provides a safe value for the plastic moment in the presence of a shear force. Horne[2] obtained a less conservative estimate by considering the plastic zones between sections A_2A_2 and A_3A_3 in Fig. 3.8, consideration being given to normal stresses acting on horizontal planes. Using the Tresca yield criterion, the plastic moment M'_p at section A_3A_3 is given by

$$\frac{M'_p}{M_p} = 1 - 0 \cdot 444 \left(\frac{F}{F_p}\right)^2 \tag{3.23}$$

provided

$$\frac{F}{F_p} \leqslant 0 \cdot 792 \tag{3.24}$$

These treatments show a reduction in the plastic moment due to shear forces. If the section A_3A_3 in Fig. 3.8 is constrained to remain plane, the moment of resistance at this section may actually exceed M_p for a range of values of F/F_p because of restraint on plastic deformation (Green[1]). However, lower bound solutions such as Eqn (3.21) or (3.23) are usually adopted if it is considered necessary to allow for the effect of shear forces on plastic hinge values, whatever the loading and support conditions may be.

Effect of shear force in I-sections

While the reduction in plastic moment given either by Eqn (3.21) or by Eqn (3.23) for rectangular sections is almost invariably negligible, cases arise in I-sections when the reduction is appreciable. Considering the section in Fig. 3.11(a), a safe (high) estimate of the reduction in plastic moment due to shear may be obtained by assuming the normal and shear force distributions shown in Figs 3.11(b) and (c). The shear stresses are assumed to be confined to the web, which has a depth $d' = d - 2t_f$ and thickness t_w. Let

$$M_{pw} = \left\{\frac{(d')^2 \, t_w}{4}\right\} \sigma_y$$

$$F_{pw} = d' t_w \tau_y$$

Then, if M_p is the plastic moment of the section without shear force, Eqn (3.21)

6—P.T.S.

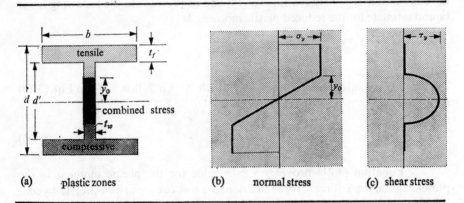

(a) plastic zones (b) normal stress (c) shear stress

Figure 3.11 Estimation of plastic moment in I-section in presence of shear force when $F/F_{pw} \leqslant \frac{2}{3}$ (see Eqn. 3.25)

may be applied to the web to determine the reduction in M_p due to a shear force F. The reduced plastic moment M'_p becomes

$$M'_p = M_p - \frac{3}{4} \left(\frac{F}{F_{pw}}\right)^2 M_{pw} \tag{3.25}$$

provided

$$\frac{F}{F_{pw}} \leqslant \frac{2}{3} \tag{3.26}$$

This solution applies as long as the plastic zones in pure tension and compression penetrate into the web. This no longer occurs when $F/F_{pw} > \frac{2}{3}$, and the stress distributions in Fig. 3.12 may then be postulated. The shear stress varies parabolically from τ_y on the central axis to τ_1 at the top and bottom of the web, where the normal stress is $\pm \sigma_1$. If the yield criterion is just satisfied

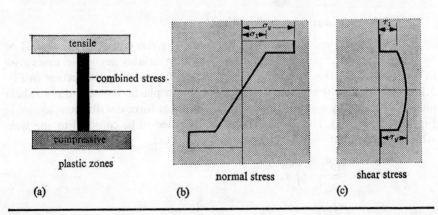

(a) plastic zones (b) normal stress (c) shear stress

Figure 3.12 Estimation of plastic moment in I-section in presence of shear force when $F/F_{pw} \geqslant \frac{2}{3}$ (see Eqn. 3.29)

under stresses σ_1, τ_1, it follows from Eqn (3.16) that

$$\left(\frac{\sigma_1}{\sigma_y}\right)^2 + \left(\frac{\tau_1}{\tau_y}\right)^2 = 1 \tag{3.27}$$

The shear force becomes, in terms of τ_y and τ_1,

$$F = d't_w\,\tau_1 + d't_w\left\{\tfrac{2}{3}\left(\tau_y - \tau_1\right)\right\}$$

whence $\dfrac{F}{F_{pw}} = \tfrac{1}{3}\left(2 + \dfrac{\tau_1}{\tau_y}\right)$ \hfill (3.28)

The moment of resistance of the section is

$$M'_p = M_p - M_{pw} + \left\{\frac{(d')^2 t_w}{6}\right\}\sigma_1$$

Using Eqns (3.27) and (3.28), this can be expressed as

$$M'_p = M_p - \left[1 - \frac{2}{3}\sqrt{\left\{1 - \left(3\frac{F}{F_{pw}} - 2\right)^2\right\}}\right]M_{pw} \tag{3.29}$$

$$\frac{2}{3} \leqslant \frac{F}{F_{pw}} \leqslant 1 \tag{3.30}$$

The reduction in plastic moment as given by Eqns (3.25) and (3.29) [expressed in the form $(M_p - M'_p)/M_{pw}$] varies with F/F_{pw} as shown by curve 1 in Fig. 3.13. When $F = F_{pw}$, the capacity of the section in pure shear has been

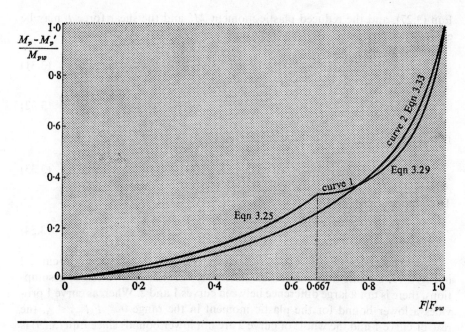

Figure 3.13 Effect of shear force F on plastic moments of resistance of I-sections

reached and unrestrained plastic deformation will take place, even if the applied moment is less than $M_p - M_{pw}$, and therefore incapable of producing fully plastic flanges.

In an alternative treatment by Heyman and Dutton[5] the shear stress is assumed at all stages to be uniform throughout the depth of the web, giving the stress distributions shown in Fig. 3.14. The stresses σ_1 and τ_1 are again related by

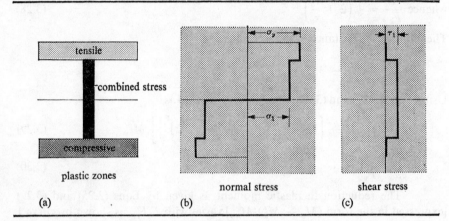

Figure 3.14 Estimation of plastic moment in I-section in presence of shear force by Heyman and Dutton's method (see Eqn. 3.33)

Eqn (3.27), and the reduced plastic moment M_p' and the shear force F may be expressed in terms of σ_1 and τ_1 as follows.

$$M_p' = M_p - \left(\frac{\sigma_y - \sigma_1}{\sigma_y}\right) M_{pw} \tag{3.31}$$

$$F = \frac{\tau_1}{\tau_y} F_{pw} \tag{3.32}$$

Eliminating σ_1 and τ_1 from Eqns (3.27), (3.31), and (3.32),

$$M_p' = M_p - \left[1 - \sqrt{\left\{1 - \left(\frac{F}{F_{pw}}\right)^2\right\}}\right] M_{pw} \tag{3.33}$$

This applies, provided

$$\frac{F}{F_{pw}} \leqslant 1 \tag{3.34}$$

The values of $(M_p - M_p')/M_{pw}$ given by Eqn (3.33) are shown by curve 2 in Fig. 3.13. Despite the considerable difference between the two sets of assumptions, there is not a large difference between curves 1 and 2. Whereas curve 1 provides a lower bound for the plastic moment in the range $0 \leqslant F/F_{pw} \leqslant \frac{2}{3}$, the rest of curve 1 and the whole of curve 2 are not lower bounds since Eqn (3.29) is derived without investigating equilibrium conditions at the junctions of web and flanges, while the solution of Heyman and Dutton [Eqn (3.33)] fails to satisfy equilibrium conditions throughout the web. However, experimental evidence (Baker, Horne, and Heyman[6]) suggests that curve 2 (which has the advantage of being given by a single formula throughout the range $0 \leqslant F/F_{pw} \leqslant 1$) is satisfactory for practical purposes.

3.4 Plastic moments under shear force and axial load

No complete solution to this problem has been derived, but approximate expressions suitable for use in design may be derived using the approach of Section 3.3 (see Horne[7]). If the I-section in Fig. 3.15(a) has a plastic moment

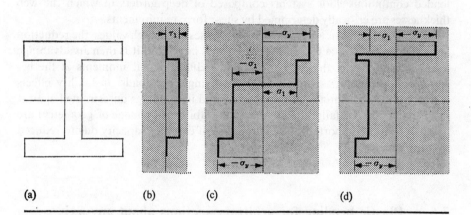

(a)   (b)   (c)   (d)

Figure 3.15 (a) Stress distributions in I-sections for the estimation of plastic moments in the presence of shear force and axial thrust; (b) Shear stress; (c) Normal stress–neutral axis in web; (d) Normal stress–neutral axis in flange

M'_p in the presence of a shear force F and axial thrust P, it may be assumed that the shear is distributed down the web as a uniform stress $\tau_1 = F/d't_w$. The longitudinal stress is then $\pm\sigma_1 = +\sigma_y \sqrt{\{1 - (\tau_1/\tau_y)^2\}}$ in the web [see Eqn 3.27)] and $\pm\sigma_y$ in the flanges. When $P < d't_w\sigma_1$, the distribution of longitudinal stress is as shown in Fig. 3.15(c), and when $P > d't_w\sigma_1$, it is as shown in Fig. 3.15(d).

The calculation of M'_p becomes somewhat tedious, particularly for the distribution in Fig. 3.15(d), but fortunately shear forces are infrequently sufficiently high to cause significant reductions in plastic moment. As a practical guide, the effect of axial load on the plastic moment can usually be neglected when the mean axial stress is less than $0 \cdot 1\sigma_y$ (and invariably when it is less than $0 \cdot 05\sigma_y$), while the effect of shear force is usually negligible when the mean shear stress in the web is less than $0 \cdot 5\tau_y$ (and invariably when it is less than $0 \cdot 25\tau_y$).

3.5   Calculation of failure loads allowing for shear force and axial thrust

The effect that shear forces and axial thrusts in individual members have on the *rigid–plastic* failure loads of low-rise rigid frames is invariably small. Hence the mode of collapse and the approximate failure load are first determined ignoring reductions in plastic moments, and approximate values are thereby derived for the shear forces and axial thrusts at plastic hinges. The reduced values of the plastic hinge moments may then be calculated, and the failure load re-calculated with these reduced values. If desired the new values of shear force and axial thrust may be used to recalculate the reduced plastic moments and the process may be repeated, but this will usually be found to make negligible difference. It is also scarcely ever found that the revised plastic moments modify the collapse mechanism, although this is a possibility in the design of heavily loaded continuous floor systems composed of deep girders in which the web thicknesses are critically determined by shear force requirements.

In buildings in which mean axial stresses reach high values, the reduction in plastic moments due to axial load becomes large, and it is then an advantage to make estimates of the axial forces and reduced plastic moments at the beginning of the analysis. In all cases involving axial loads, instability effects (Chapter 6) are potentially of importance, and in buildings unrestrained against sway, reductions of failure loads due to instability (or change of geometry) are liable to be more important than reductions of carrying capacity due to reduced plastic moments.

3.6   Plastic bending for arbitrary sections about any axis

Let RST in Fig. 3.16(a) represent the outline of any cross section of area A, the centroid being at G. If plastic flexure takes place under zero axial load, with neutral axis AA, then AA must be an equal area axis, dividing the cross section into areas $A/2$ with centroids at g_1 and g_2. The centroid G lies on $g_1 g_2$ and bisects it. The plastic moment will be of magnitude $(g_1 g_2) A \sigma_y / 2$ and must act in the plane containing $g_1 g_2$ and the longitudinal axis, not in general perpendicular to the neutral axis AA. In this respect, therefore, a section behaves qualitatively at full plasticity in the same fashion as in the elastic range when the section is not bent about a principal axis.

We now consider an adjacent equal area axis A'A', the centroids of the half areas moving to g_1' and g_2'. The displacements $g_1 g_1' = g_2 g_2'$ are due to the transfer of each of the shaded areas from one half of the section to the other. To the first order of small quantities, therefore, as the angle between AA and A'A' becomes small, $g_1 g_1'$ and $g_2 g_2'$ become parallel to AA (see Brown[8]). Hence the locus of g_1 and g_2 is a closed curve with radial symmetry about the centroid

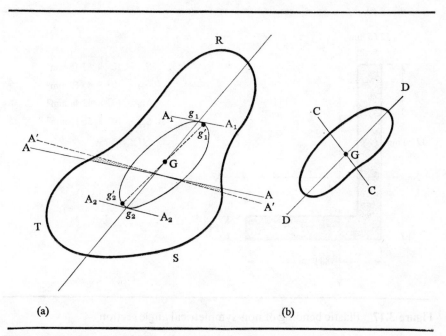

(a)   (b)

Figure 3.16   Plastic bending of member of arbitrary cross section

G, and is readily shown to be concave inwards. Since there must at least be two diameters CC and DD [Fig. 3.16(b)] of the locus of g_1 and g_2 which are normal to the locus, there must always be at least two axes of plastic bending of a section which are perpendicular to the planes containing the corresponding bending moments. These longitudinal planes of a member may be called the *principal planes* for plastic bending. For a member with at least one axis of symmetry, at least two principal planes are mutually perpendicular and coincide with the principal axes of inertia of the cross section, but for a member with no axis of symmetry, principal planes are not in general mutually at right angles and are not coincident with the principal axes of inertia.

      Any problems involving the bending of an unsymmetrical section, or of a symmetrical section about an inclined axis, are best solved by specifying the plane of flexure and calculating the total bending moments from the resulting areas into which the section is divided by the corresponding equal area axis. Thus if the angle section in Fig. 3.17 bends about the axis XX (that is, plane of flexure YY), the resulting bending moment has components $S_{xx}\sigma_y$ about XX and $S_{xy}\sigma_y$ about YY, where $S_{xx} = 10^3 \times 87\cdot16$ mm^3 and $S_{xy} = 10^3 \times 42\cdot44$ mm^3 (see Horne[9] and Burnett[10]). On the other hand, if bending takes place about axis YY, the bending moment components are $S_{yy}\sigma_y$ about YY and $S_{yx}\sigma_y$ about XX, where $S_{yy} = 10^3 \times 45\cdot69$ mm^3 and $S_{yx} = 10^3 \times 56\cdot13$ mm^3. It is to be noted that $S_{xy} \neq S_{yx}$.

      Any member bent plastically in a plane which is not an axis of symmetry is liable to twist unless it is restrained from doing so at frequent intervals. An

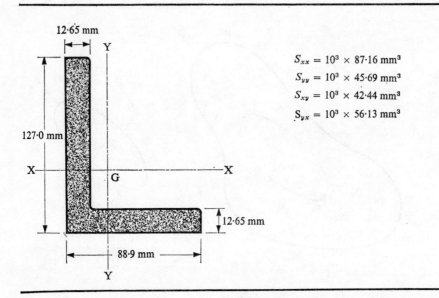

Figure 3.17 Plastic bending of non-symmetrical angle section

important practical case is that of angle or channel sections used as purlins or sheeting rails. Plastic theory may be applied provided there is no local buckling and provided the connections to the cladding are sufficient to prevent serious twisting of the member. In such cases the applied loads are resisted by bending moments $S_{xx}\sigma_y$ or $S_{yy}\sigma_y$ as previously defined, while moments $S_{xy}\sigma_y$ or $S_{yx}\sigma_y$, perpendicular to the plane of flexure, are absorbed by shearing forces transmitted in the cladding.

3.7   Design of joints

The usually stated assumption regarding joints in structures for which plastic collapse loads are being calculated is that they are 'rigid'. Strictly speaking, this requirement refers to a condition necessary for the application of elastic theory. In order that a structure may attain an ultimate load equal to the theoretical plastic collapse load, the joints need not be completely 'rigid', but they must satisfy the two following requirements.

(a)   The joint must be capable of transmitting the appropriate plastic moment or moments.

(b)   The relative rotations of the ends of members at the joint when these plastic moments are transmitted must be within a small factor (say two or three) of the relative rotations that would arise owing to elastic deformations in the structure *if that joint were rigid*.

The second requirement means that, while the joint need not be completely rigid, the degree of rigidity must be sufficient to ensure that for a reasonable range of deformation of the structure as a whole, the bending moments can become redistributed in accordance with the assumptions of plastic theory.

Beam to column joints

Although welding necessarily features prominently in the design of full-strength joints, site joints need not themselves be welded. The most common joint in a steel frame is that between an I-beam and the flange of a column. The beam may either be welded directly to the column (Fig. 3.18) or to an end plate

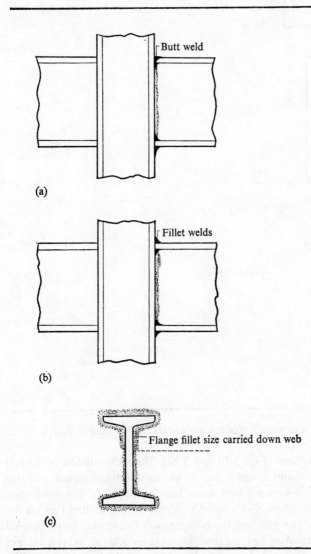

Figure 3.18 Welding for full-strength beam connections

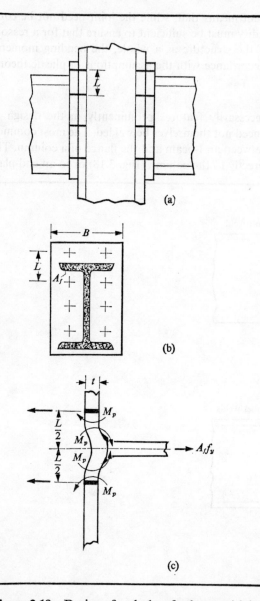

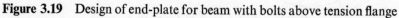

Figure 3.19   Design of end-plate for beam with bolts above tension flange

which is bolted to the column (Figs 3.19 and 3.20). The transmission of the full
strength of the upper (tension) flange is usually the most critical aspect, and this
may be achieved by a full-strength butt weld. For *in situ* joints, this involves a
single vee preparation to allow down-hand welding, with a sealing run on the
underside [Fig. 3.18(a)]. The web and the compression flange may be completed
by fillet welds. Fillet welds may be used throughout [Figs 3.18(b) and (c)] and are
found to be satisfactory provided the combined throat thickness of the fillet

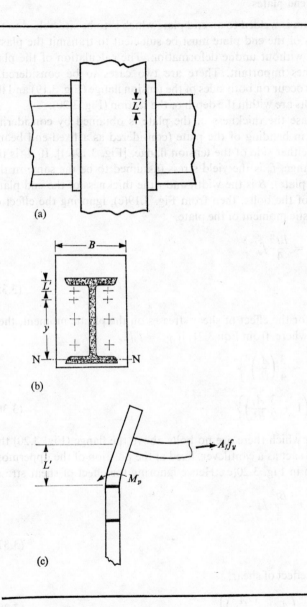

(a)

(b)

(c)

Figure 3.20 Design of end-plate for beam with all bolts between flanges

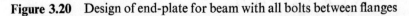

welds is at least equal to the thickness of the metal being joined (that is, throat thickness = $t_f/2$ for equal welds round the flanges, where t_f is the flange thickness of the beam). It is essential to continue the heavier flange weld round past the root fillet and down the web for not less than 50 mm to eliminate the possibility of cracking at the root fillet [see Fig. 3.18(c)].[11–13]

Thickness of end plates

When the beam is welded to an end plate which is to be site bolted to the column, the thickness of the end plate must be sufficient to transmit the plastic moment of the beam without undue deformation. The calculation of the plate thickness thus becomes important. There are two cases to be considered,[14] one in which the bolts occur on both sides of the tension flange (Fig. 3.19) and the other where all the bolts are within the depth of the section (Fig. 3.20).

In the first case the thickness of the plate is obtained by considering the ultimate strength in bending of the plate (considered as a fixed-end beam) between the bolts on either side of the tension flange, [Fig. 3.19(c)]. If A_f is the area of the tension flange, f_y is the yield stress (assumed to be the same in the flange and in the end plate), B is the width and t the thickness of the end plate, and L is the spacing of the bolts, then from Fig. 3.19(c), ignoring the effect of shear stress on the plastic moment of the plate,

$$\frac{A_f f_y L}{8} = M_p = \frac{Bt^2}{4} f_y$$

or $\qquad t = \sqrt{\dfrac{A_f L}{2B}}$ \hfill (3.35)

If allowance is made for the effect of shear stresses on the plastic moment, then M_p is replaced by M_p' where, from Eqn 3.21, if $\gamma_g = f_y/2$,

$$M_p' = M_p \left\{ 1 - \frac{3}{4} \left(\frac{A_f}{Bt} \right)^2 \right\}$$

Hence $\quad t = \sqrt{\left\{ \dfrac{A_f L}{2B} \left(1 + \dfrac{3}{2} \dfrac{A_f}{BL} \right) \right\}}$ \hfill (3.36)

In the case in which there are no bolts above the flange (Fig. 3.20) the end plate is assumed to act as a cantilever, fixed at the position of the uppermost row of bolts as shown in Fig. 3.20(c). Hence ignoring the effect of shear stress,

$$A_f f_y L' = M_p = \frac{Bt^2}{4} f_y$$

or $\qquad t = 2 \sqrt{\dfrac{A_f L'}{B}}$ \hfill (3.37)

Taking account of the effect of shear,

$$t = 2 \sqrt{\left\{ \frac{A_f L'}{B} \left(1 + \frac{3}{4} \frac{A_f}{BL'} \right) \right\}}$$ \hfill (3.38)

Size of bolts

When the bolts are equally spaced about the tension flange, the net area A_b required per bolt is obtained simply by equating the flange force $A_f f_y$ to the total bolt force $N A_b f_y'$, where N is the number of bolts and f_y' is the yield stress in the bolts. This would be the approach for sizing the four bolts round the upper (tension) flange in Fig. 3.19. If there are no bolts above the tension flange (Fig. 3.20), the procedure is to assume that the neutral axis for bending action on the connection interface is NN, coincident with the compression flange. Moments are then taken about this neutral axis, so that if M is the moment of resistance and M_p is the plastic moment of the beam,

$$M_p < M = \sum y A_b f_y' \qquad (3.39)$$

where y is the lever arm of a typical bolt.

Strength of column web

Stiffeners may be required in the column to prevent local bending of the column flange accompanied by buckling or tearing of the column web. It has been proposed[15] that stiffeners should be provided if the thickness of the column web t_w is less than $A_f/(t_f + 5k)$, where A_f is the area of the flange of the beam, t_f is the mean flange thickness of the column, and k is the distance from the outer face of the column to the point in the web where the root fillet ends [Fig. 3.21(b)].

When beams of the same order of size are rigidly connected to each side of the column, [Fig. 3.21(a)], there is no danger of failure due to shear in the column web. With highly unequal beams, or when a beam is connected

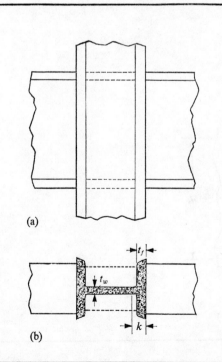

(a)

(b)

Figure 3.21 Provision of stiffeners to column webs

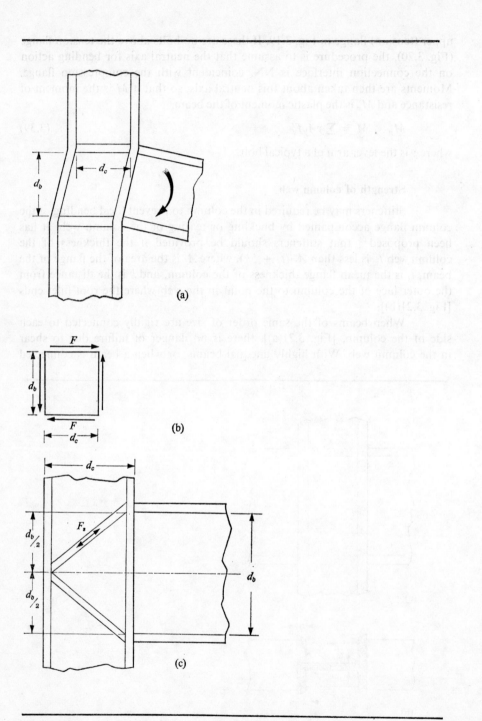

(a)

(b)

(c)

Figure 3.22 Ultimate strength and stiffening of unbalanced beam to column connections

to one side only, failure may occur by shearing in the column web as shown in Fig. 3.22(a). The maximum unbalanced moment M may be calculated by considering the maximum shear force F acting on the enclosed area of web, [Fig. 3.22(b)]. If the dimensions of the enclosed area are $d_b \times d_c$ as shown, and the thickness of the web is t_w, then $F = d_c t_w f_y/2$ (assuming a yield stress in shear of $f_y/2$), and hence

$$M = Fd_b = \frac{d_b d_c t_w f_y}{2}$$

If there is a beam on one side only of the joint and Z_p is the plastic modulus, we must have $M \geqslant M_p = Z_p f_y$, so that

$$Z_p \leqslant \tfrac{1}{2} d_b d_c t_w \tag{3.40}$$

When Z_p exceeds the above limit, stiffeners must be provided. If 'k' stiffeners each of area A_s are welded between column flanges on each side of the column as shown in Fig. 3.22(c), the total stiffener force F_s is $F_s = 2A_s f_y$, so that the horizontal component F' is

$$F' = 2A_s f_y \frac{d_c}{\sqrt{(d_c^2 + 0{\cdot}25d_b^2)}}$$

The moment taken by the stiffeners is $F'd_b$, so that the total moment of resistance of the joint becomes

$$M = \frac{d_b d_c t_w f_y}{2} + \frac{2A_s f_y d_b d_c}{\sqrt{(d_c^2 + 0{\cdot}25d_b^2)}}$$

Since $M \geqslant Z_p f_y$, it follows that

$$A_s \geqslant \left(Z_p - \frac{d_b d_c t_w}{2}\right) \frac{\sqrt{(d_c^2 + 0{\cdot}25d_b^2)}}{2d_b d_c} \tag{3.41}$$

It is interesting to see from this what a ready application plastic theory has to the design of connections. It is in fact plastic theory that is effectively used in many so-called 'elastic' methods of structural design. Stress concentrations introduced by the complexity of structural details may be very high, but, provided the material is ductile, have little significance when calculating ultimate strengths.

Bibliography

1   Green, A. P. A theory of the plastic yielding due to bending of cantilevers and fixed-ended beams. *J. Mech. Phys. Solids*, 1954, **3**, 143.

2   Horne, M. R. The plastic theory of bending of mild steel beams with particular reference to the effect of shear forces. *Proc. R. Soc.*, 1951, **207**[A], 216.

3   Neal, B. G. Limit load of a cantilever in plane stress. In Heyman, J. and
       F. A. Leckie (Eds.), *Engineering Plasticity*, Cambridge University Press,
       London, 1958, 473.

4   Neal, B. G. *The Plastic Methods of Structural Analysis*, Chapman and
       Hall, London, 1956.

5   Heyman, J. and V. L. Dutton. Plastic design of plate girders with
       unstiffened webs. *Welding and Metal Fabrication*, 1954, **22**, 265.

6   Baker, J. F., M. R. Horne, and J. Heyman. *The Steel Skeleton*, Vol. II,
       Cambridge University Press, London, 1956.

7   Horne, M. R. The full plastic moments of sections subjected to shear
       force and axial loads. *British Welding Journal*, 1958, **5**, 170.

8   Brown, E. H. Plastic asymmetrical bending of beams. *Int. J. Mech. Sci.*,
       1967, **9**, 77.

9   Horne, M. R. *The Plastic Properties of Rolled Sections*, Pamphlet
       T.31A, British Welding Research Association, 1959.

10   Burnett, N., L. G. Johnson, L. J. Morris, A. L. Randall, and C. P.
       Thompson. *Plastic Design*. British Constructional Steelwork Asso-
       ciation, Publication No. 28, 1965.

11   Johnson, L. G. Tests on welded connections between I-section beams
       and stanchions. *British Welding Journal*, 1959, **6**, 38.

12   Johnson, L. G. Further tests on welded connections between I-section
       beams and columns. *British Welding Journal*, 1959, **6**, 367.

13   Morris, L. J. Tests on fillet welded rigid connections between universal
       beams and stanchions in high strength steel to BS 968:1962, in *Welding
       of BS968:1962 Steel*, The Institute of Welding, 1965.

14   Sherbourne, A. N. Bolted beam to column connections. *Structural
       Engineer*, 1961, **39**, 203.

15   Khabbaz, R. N., J. D. Graham, and A. N. Sherbourne. Welded interior
       beam to column connections. *W.R.C. Bulletin Series*, No. 63, Welding
       Research Council, 1960.

Problems

3.1   The I-section member in Fig. 3.4(a) is bent about an axis in the plane of
       the web. Obtain expressions for the plastic modulus Z_p' under an axial
       thrust $P = nA\sigma_y$, where A is the cross-sectional area and σ_y is the yield
       stress.

3.2   An I-section is of total depth 0·50 m, flange width 0·20 m, flange thick-
       ness 15 mm, and web thickness 10 mm. If the yield stress is 200×10^6
       N/m^2, find from first principles the plastic moments under an axial
       thrust of (a) 500 kN, (b) 1500 kN when bent about the major axis.
       Repeat for bending about the minor axis.

3.3   The cross section of a beam is an isosceles triangle of base b and depth d. The beam has its plane of symmetry vertical with the apex of the cross section pointing downwards and is subjected to a sagging bending moment. Calculate the plastic moment $M_p' = Z_p' \sigma_y$ under an axial thrust $P = nbd\sigma_y/2$, where σ_y is the yield stress. Find the value of n for which Z_p' is a maximum and the corresponding value of Z_p'. Moments are measured about the centroidal axis.

3.4   A fixed-base portal frame ABCD, fixed at A and D, consists of a horizontal rigid beam BC supported on vertical columns AB, CD of height H, distance L apart. The columns are of rectangular cross section of width b, their depth in the plane of the frame being d, and the material has a yield stress σ_y. The beam is subjected to a horizontal load W in the plane of the frame. Calculate W when plastic collapse occurs, allowing for the effect on the plastic moments in the columns of axial thrust but neglecting the effect of shear. Assume that no plastic deformation occurs in the beam.

3.5   A channel section is of overall dimensions $d \times 2d$, the flanges being of width d and thickness $2t$ and the web of depth $2d$ and thickness t. The yield stress is σ_y and t is small compared with d. The channel is simply supported with its flanges in vertical planes, the web being uppermost. Calculate relationships between the axial force P (thrust positive) and bending moment M (sagging moment positive) to produce full plasticity over the range of P from $-6td\sigma_y$ to $6td\sigma_y$ for both positive and negative values of M. The moment is measured about the centroidal axis. Find the value of P which gives M its maximum positive value, and the corresponding value of M.

3.6   A symmetrical I-beam of length 10 m has a plastic modulus of 1.21×10^{-3} m^3, a web thickness of 9.1 mm and a web depth of 333 mm. The material has a yield stress σ_y in pure tension or compression of 250×10^6 N/m^2 and a yield stress τ_y in pure shear of 140×10^6 N/m^2. Assuming that the shear stress is distributed uniformly down the web, and that the beam is fixed-ended and carries a uniformly distributed load, estimate the plastic failure load, allowing for the effect of shear forces in the end sections. What is the percentage reduction in the carrying capacity of the beam due to shear? Assume that the combinations of direct stress σ_1 and shear stress τ_1 to produce yield are given by Eqn (3.27).

3.7   A member of rectangular cross section, width b and depth d, consists of material with yield stresses in tension of f_t and in compression of f_c. Show that the moment of resistance at full plasticity about a horizontal axis is $f_t f_c bd^2/2(f_t + f_c)$. If $f_t = (1 - \beta)f_y$ and $f_c = (1 + \beta)f_y$ so that f_y is the mean yield stress, show that the error in calculating the plastic moment by assuming f_y to be the yield stress in tension and compression is $(100\beta^2)\%$.

Chapter 4
Minimum weight design

4.1   Introduction

The previous chapters have shown how the collapse load factor of a given structure may be derived, that is, the structure has been *analysed* in relation to plastic collapse. The process of *design* is one in which the load factor is required to have a given minimum value, and the plastic moments of the various members of the structure are required. Design may be carried out in a limited sense, using any of the analytical methods that have been described, merely by assigning ratios to the plastic moments of resistance of the members. This process was used in designing the portal frame in Fig. 2.1, in that a uniform section was assumed throughout the frame. In the more general case, however, the members do not necessarily have preassigned ratios of strength, and the design process effectively involves choosing what the ratios are to be.

When a design is required for a single set of loads, *any* bending moment distribution satisfying the equilibrium and yield conditions constitutes a possible basis for design. Suppose the continuous beam in Fig. 4.1(a) is to be designed at unit load factor for the loads shown. The corresponding free bending moments are given by A*b*C*d*E*f*G in Fig. 4.1(b). If the central span CE is to be given the minimum section possible, the reactant moment line must be *ce* as shown, giving a plastic moment of 4 units. The reactant line is then determined also for the other spans, giving required plastic moments of 10 and 13 for spans AC and EG respectively. Another possible solution is shown in Fig. 4.1(c). An infinite number of solutions actually exists for the design of this continuous beam, and factors other than strength may be introduced to decide what is, in some sense, the best design. Examples of such factors are limiting deflections, minimum total weight, availability of sections, convenience of fabrication, and minimum total cost. The two latter considerations would almost certainly, in the case of the beam in Fig. 4.1(a), lead to the use of a uniform section throughout with a plastic moment of 10 units [Fig. 4.1(c)]. In more complicated cases it becomes less easy to see what is likely to be the best practical solution, and it is therefore useful to pursue at any rate one criterion which is likely to be important and which is capable of formal treatment. This is the criterion of *minimal total weight*, and

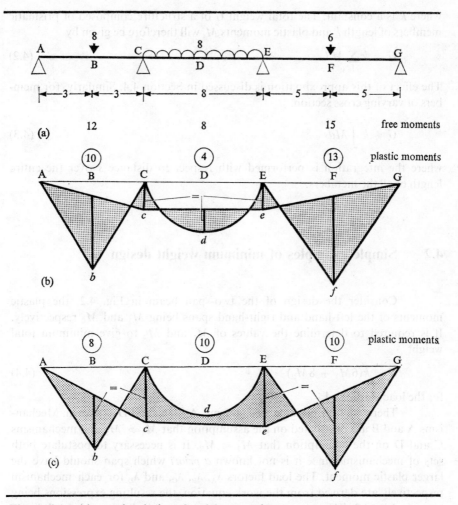

Figure 4.1 Alternative designs for three-span beam

this is the concern of the present chapter. Sections 4.2 to 4.4 deal with minimum weight design using prismatic members, and the use of members of continuously varying cross section is discussed in Section 4.5.

It is necessary to adopt some relationship between weight per unit length g and plastic moment M. For materials of given yield stress and geometrically similar section, $g \propto M^{2/3}$. When the more economic Universal Beam Sections are considered, it is found that, very closely, $g \propto M^{0.6}$, and more generally one could put $g \propto M^n$. It is found that the relative plastic moments for absolute minimum weight in a given structure are surprisingly little affected by the exact value chosen for the index n, and since solutions are much facilitated by retaining a linear relationship, it will be assumed that $n = 1$. Hence, it will be assumed that the weight per unit length is given by

$$g = kM \qquad (4.1)$$

where k is a constant. The total weight G of a structure composed of prismatic members of length l_i and plastic moments M_i will therefore be given by

$$G = k \sum_i M_i l_i \qquad (4.2)$$

The effect of this approximation is discussed in Section 4.4. Similarly, for members of varying cross section,

$$G = k \int M ds \qquad (4.3)$$

where the integration is performed with respect to distance s over the entire length of all the members.

4.2   Simple examples of minimum weight design

Consider the design of the two-span beam in Fig. 4.2, the plastic moments of the left-hand and right-hand spans being M_1 and M_2 respectively. It is required to determine the values of M_1 and M_2 to give minimum total weight

$$G = k(6M_1 + 8M_2) \qquad (4.4)$$

for the loads indicated.

There are four possible mechanisms, A, B, C, and D as shown. Mechanisms A and B are postulated on the assumption that $M_1 > M_2$ and mechanisms C and D on the assumption that $M_1 < M_2$. It is necessary to postulate both sets of mechanisms since it is not known *a priori* which span should have the larger plastic moment. The load factors λ_A, λ_B, λ_C, and λ_D for each mechanism respectively are derived from the work equation, the resulting expressions being given in Fig. 4.2. These expressions are shown graphically for unit load factor $(\lambda_A = \lambda_B = \lambda_C = \lambda_D = 1)$ by the four continuous straight lines in Fig. 4.3. Combinations of M_1 and M_2 to the right of and above these mechanism lines represent continuous beams which will not fail by the respective mechanisms below unit load factor. Hence the area bounded by these mechanism lines forms a *permissible region* in which any allowable design must lie. The boundary to the region is shown by the shading in Fig. 4.3, and involves mechanisms B, C, and D. It thus appears that the beam will not, under the given loads, fail by mechanism A whatever the ratio of M_1 to M_2.

Not only may the mechanism equations be represented by straight lines in Fig. 1.3, but so also may the total weight Eqn 4.4. The straight line *ab* represents Eqn 4.4 if $Oa/Ob = \frac{6}{8}$ and $Oc = G/k \sqrt{(6^2 + 8^2)} = G/10k$. The minimum weight possible for this continuous beam is therefore represented to scale by Oc', where $a'b'$ (parallel to *ab*) just touches the permissible region. Hence the *minimum weight design* is represented by point *d*, giving $M_1 = 0.75$, $M_2 = 3.625$,

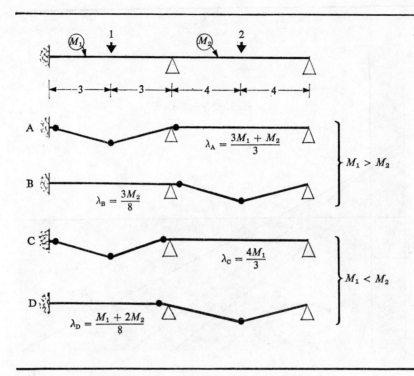

Figure 4.2 Possible design mechanisms for two-span beam

and $G = 33 \cdot 5k$. All other points within, or upon the boundary of, the permissible region give a weight greater than $33 \cdot 5k$; thus point f, the intersection of mechanism lines B and D, gives $M_1 = M_2 = 2 \cdot 67, G = 37 \cdot 3k$.

This example involves only two independent varying plastic moments, but the same graphical presentation helps to identify the nature of the problem when there are n independent plastic moments. The clue is to be found in noting that the *slope* of the weight line $a'b'$ which touches the boundary of the permissible region at d is intermediate between the slopes of the mechanism lines C and D which intersect at d. It is known that the equation of the weight line is of the form of Eqn 4.4, whereas mechanism lines C and D have respectively the equations

$$3 = 4M_1 \tag{4.5}$$

$$8 = M_1 + 2M_2 \tag{4.6}$$

The statement that $a'b'$ has a slope intermediate between those of Eqns 4.5 and 4.6 means, algebraically, that it must be possible to combine these equations with positive multipliers and thereby obtain Eqn 4.4. Thus, if Eqn 4.5 is combined with μ times Eqn 4.6, the resulting equation

$$3 + 8\mu = (4 + \mu)M_1 + 2\mu M_2 \tag{4.7}$$

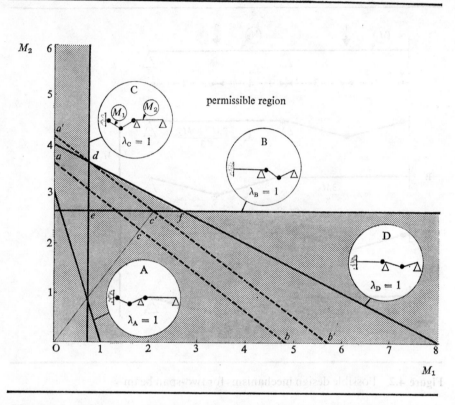

Figure 4.3   Minimum weight interaction diagram for beam in Fig. 4.2.

is identical in slope with Eqn 4.4 if $\mu = 8$. The addition of mechanism C to eight times mechanism D leads to the mechanism in Fig. 4.4. This mechanism is seen to be such that, in the work equation (as shown), *the total hinge rotation $\sum \theta$ associated with any plastic moment of resistance M is proportional to the total lengths of members with that plastic moment.* This may be extended to any structure with any number of independent plastic moments over which there is freedom of

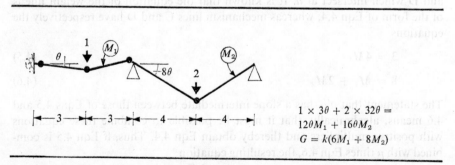

$$1 \times 3\theta + 2 \times 32\theta = 12\theta M_1 + 16\theta M_2$$
$$G = k(6M_1 + 8M_2)$$

Figure 4.4   Minimum weight mechanism for beam in Fig. 4.2

choice in the design process. If such a mechanism can be found, and if the associated plastic moments and distribution of bending moments allow the *equilibrium and yield conditions* to be satisfied (that is, if the design is not outside the permissible region), then the structure has the minimum possible weight.

It is to be noted that the design achieved is not necessarily unique. If the weight line is tangential to the permissible region over a finite range, a range of minimum weight designs is possible. This is true, for example, for the portal frame in Fig. 4.5(a), for which the permissible region and minimum weight line are as shown in Fig. 4.6. The weight line

$$G = k(2M_1 + 2M_2) \qquad (4.8)$$

is parallel to the mechanism line

$$3\theta = 2\theta M_1 + 2\theta M_2 \qquad (4.9)$$

for the mechanism in Fig. 4.5(b). Since the latter is found to form part of the boundary to the permissible region, the minimum weight line coincides with this mechanism line over the finite distance *ab* as shown in Fig. 4.6. The solution

$$\tfrac{3}{2} = M_1 + M_2$$

giving $G = 3k$ is valid for any point between *a* and *b*, that is, from $M_1 = M_2 = \tfrac{3}{4}$ to $M_1 = 1, M_2 = \tfrac{1}{2}$.

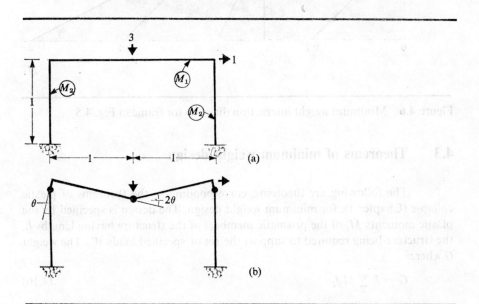

Figure 4.5 Portal frame with minimum weight mechanism

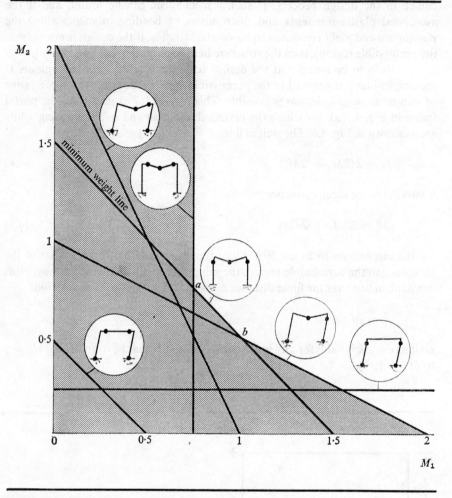

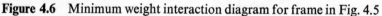

Figure 4.6 Minimum weight interaction diagram for frame in Fig. 4.5

4.3   Theorems of minimum weight design

The following are theorems, corresponding to the theorems of plastic collapse (Chapter 1), for minimum weight design. The design is specified by the plastic moments M_j of the prismatic members of the structure having lengths l_j, the structure being required to support the set of specified loads W_i. The weight G where

$$G = k \sum_j M_j l_j \qquad (4.10)$$

is required to be an absolute minimum, k being a constant. These theorems were first discussed by Foulkes.[1, 2]

Uniqueness theorem of minimum weight

Suppose it is possible to postulate a mechanism involving total positive plastic hinge rotations ϕ_j associated with the positive plastic moments M_j, together with corresponding displacements Δ_i associated with the loads W_i, such that in the work equation

$$\sum_i W_i \Delta_i = \sum_j M_j \phi_j \qquad (4.11)$$

the condition

$$\phi_j = \alpha l_j \qquad (4.12)$$

is satisfied where α is a constant. Any total plastic rotation ϕ_j is composed of individual positive rotations θ_{jk} at points h_{jk}, so that

$$\phi_j = \sum_k \theta_{jk} \qquad (4.13)$$

Then, providing the bending moments throughout the structure satisfy the equilibrium and yield conditions for plastic collapse as defined in Section 1.5, the structure is a minimum weight structure for the given loads.

Proof. Suppose any design has plastic moments M_j^*, so that the weight of the design is given by

$$G^* = k \sum_j M_j^* l_j \qquad (4.14)$$

Let the bending moment at any point h_{jk} when this design sustains the loads W_i be denoted by M_{jk}. Since the equilibrium condition is satisfied,

$$\sum_i W_i \Delta_i = \sum_j \left(\sum_k M_{jk} \theta_{jk} \right) \qquad (4.15)$$

Now the moment M_{jk} satisfies the yield condition, so that $-M_j^* \leqslant M_{jk} \leqslant M_j^*$. Hence from Eqn (4.15), and using Eqn (4.13),

$$\sum_i W_i \Delta_i \leqslant \sum_j M_j^* \phi_j \qquad (4.16)$$

Using Eqns (4.12), (4.14), and (4.16),

$$\sum_i W_i \Delta_i \leqslant \frac{\alpha}{k} G^*$$

However, from Eqns (4.10), (4.11), and (4.12),

$$\sum_i W_i \Delta_i = \frac{\alpha}{k} G$$

Hence $G^* \geqslant G$

A design thus gives the minimum weight if it satisfies the following *four* conditions.

(a)   *equilibrium* condition;
(b)   *yield* condition;
(c)   *mechanism* condition;
(d)   *plastic hinge* condition [Eqn (4.12)].

The first three conditions are identical with those for the plastic collapse of any structure, and it is the fourth condition that imposes minimum weight. Upper and lower bound theorems, analogous to those for plastic collapse, may be established in relation to the minimum weight G.

Upper bounds on minimum weight

Any design for which a set of bending moments satisfying conditions (a) and (b) is available gives an upper bound on the minimum weight.

This statement follows directly from the lower bound theorem of plastic collapse, since a structure so designed will not collapse when the loads are multiplied by a factor less than unity, and the design is therefore a safe one. The weight of the structure must therefore be greater than or equal to the minimum weight. In terms of the graphical representations in Figs 4.3 and 4.6, any design satisfying the equilibrium and yield conditions must lie within the permissible region, and cannot lie nearer to the origin than the tangent weight line.

Any design for which a set of moments satisfying conditions (a), (b), and (c) is available will just collapse under the specified loads, and therefore also gives an upper bound on the minimum weight. In terms of the graphical representation, such a design lies on the boundary of the permissible region.

Lower bounds on minimum weight

Any design satisfying conditions (c) and (d) for minimum weight provides a lower bound on the minimum weight. For example, a weight line (parallel to ab) may be drawn through e in Fig. 4.3, representing a positive combination of mechanisms B and C, and the corresponding weight is a lower bound on the minimum weight. Thus, combining the work equations for mechanisms B and C for $\lambda_B = \lambda_C = 1$ (Fig. 4.2),

$$8 + 3\mu = 4\mu M_1 + 3M_2$$

giving a line parallel to the weight Eqn (4.4) when $4\mu/3 = \frac{6}{8}$, that is $\mu = \frac{9}{16}$. The combined mechanism (Fig. 4.7) satisfies both conditions (c) and (d), but since $M_1 = 0.75$ and $M_2 = 2.67$, the equilibrium condition is certainly not satisfied over the central support. Nevertheless, the resulting weight ($G = 25.83k$) is a lower bound on the minimum weight ($G = 33.5k$).

Proof. Suppose plastic moments M_j^* with corresponding total hinge rotations $\phi_j^* = |\sum_k \theta_{jk}^*|$ (where individual hinge rotations θ_{jk} occur at points

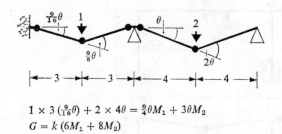

$$1 \times 3 \left(\tfrac{9}{16}\theta\right) + 2 \times 4\theta = \tfrac{9}{4}\theta M_1 + 3\theta M_2$$
$$G = k\,(6M_1 + 8M_2)$$

Figure 4.7 Combined mechanism for point *e* in Fig. 4.3

h_{jk}^*) may be postulated such that the total hinge rotations satisfy the condition

$$\phi_j^* = \alpha l_j \tag{4.17}$$

and the corresponding displacements Δ_i^* satisfy

$$\sum_i W_i \Delta_i^* = \sum_j M_j^* \phi_j^* \tag{4.18}$$

The weight of the structure is given by

$$G^* = k \sum_j M_j^* l_j = \frac{k}{\alpha} \sum_i W_i \Delta_i^* \tag{4.19}$$

Suppose the minimum weight structure has plastic moments M_j and total weight G. Let the moments at points h_{jk}^* in the minimum weight structure when it supports the loads W_i be M_{jk} where, since the yield condition is satisfied, $-M_j \leqslant M_{jk} \leqslant M_j$. Since also the equilibrium condition is satisfied, by considering the rotations θ_{jk}^* and corresponding displacements Δ_i^*,

$$\sum_i W_i \Delta_i^* = \sum_j \left(\sum_k M_{jk} \theta_{jk}^*\right) \leqslant \sum_j M_j \phi_j^* \tag{4.20}$$

Hence, from Eqns (4.10), (4.17), and (4.20),

$$G = k \sum_j M_j l_j = \frac{k}{\alpha} \sum_j M_j \phi_j^* \geqslant \frac{k}{\alpha} \sum_i W_i \Delta_i^* \tag{4.21}$$

whence, from Eqn (4.19), $G^* \leqslant G$.

The use of the minimum weight theorems to obtain minimum weight structures is illustrated in the following section.

4.4   Derivation of minimum weight frames

Two-bay frame

It is required to find the minimum weight two-bay frame in Fig. 4.8(a), loaded as shown. The beams BD and DG are required to have a common

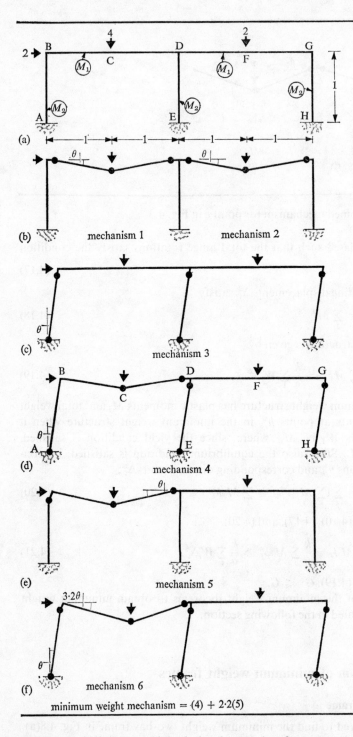

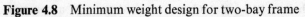

minimum weight mechanism = (4) + 2·2(5)

Figure 4.8   Minimum weight design for two-bay frame

plastic moment M_1 and the three columns a common plastic moment M_2. The total weight of the structure is then (omitting the constant multiplying factor),

$$G = 4M_1 + 3M_2 \qquad (4.22)$$

Lower bound estimates of M_1 and M_2 may be obtained from the elementary mechanisms shown in Figs 4.8(b) and (c). Mechanism 1 in Fig. 4.8(b) gives $M_1 = 1$ and mechanism 2 gives $M_1 = \frac{1}{2}$, while mechanism 3 in Fig. 4.8(c) gives $M_2 = \frac{1}{3}$. These mechanism lines are shown in Fig. 4.9. The larger lower bound on a minimum weight is represented by point a ($M_1 = 1$, $M_2 = \frac{1}{3}$), whence $G_{\min} \geqslant 4(1) + 3(\frac{1}{3}) = 5$.

The next step is to maintain M_1 and M_2 in the same ratio, but to increase their values until a point on the boundary of the permissible region is obtained. This is represented in Fig. 4.9 by proceeding along the straight line

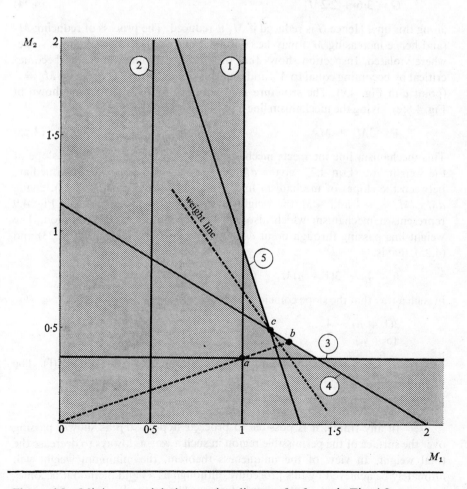

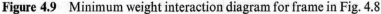

Figure 4.9 Minimum weight interaction diagram for frame in Fig. 4.8

Oa, with $M_1 = 3M_2$. The values of M_1 and M_2 required to support the given loads at unit load factor may be determined by any of the methods described in Chapter 2, whence the collapse mechanism becomes as shown in Fig. 4.8(d) and $M_1 = 3M_2 = \frac{9}{7}$. These moments are represented by point b in Fig. 4.9, and, since the upper bound conditions are satisfied, $G_{\min} \leqslant 4(\frac{9}{7}) + 3(\frac{3}{7}) = 6\cdot43$. At this stage, therefore, it is known that

$$5 \leqslant G_{\min} \leqslant 6\cdot43$$

The mechanism line passing through b in Fig. 4.9 may be obtained from the work equation for the mechanism in Fig. 4.8(d), and is

$$6 = 3M_1 + 5M_2 \tag{4.23}$$

Consider now the effect of moving *from* point b along the mechanism line. Eliminating M_2 between Eqns (4.22) and (4.23),

$$G = 3\cdot6 + 2\cdot2M_1 \tag{4.24}$$

along this line. Hence G is reduced if M_1 is reduced. The process of reducing M_1 (and hence increasing M_2) may be continued until the yield condition is somewhere violated. Inspection shows that the bending moment at B first becomes critical by becoming equal to M_2 and that this occurs when $M_1 = \frac{7}{6}$ and $M_2 = \frac{1}{2}$ (point c in Fig. 4.9). The structure may now fail by the mechanism shown in Fig. 4.8(e), giving the mechanism line 5 in Fig. 4.9, namely

$$4 = 3M_1 + M_2 \tag{4.25}$$

This mechanism line intersects mechanism line 4 at point c. Since the slope of the weight line (Eqn 4.22, giving $dM_2/dM_1 = -\frac{4}{3}$) is seen to be intermediate between the slopes of mechanism lines 4 and 5 [Eqns (4.23) and (4.25), giving $dM_2/dM_1 = -\frac{3}{5}$ and -3], the weight line passing through point c in Fig. 4.9 represents a mechanism which also satisfies the plastic hinge condition. The weight line passing through point c is obtained by combining Eqns (4.23) and (4.25), that is,

$$6 + 4\mu = 3(1 + \mu)M_1 + (5 + \mu)M_2$$

in such a way that the slope coincides with that of Eqn (4.22). Hence

$$\frac{3(1 + \mu)}{(5 + \mu)} = \frac{4}{3}$$

that is, $\mu = 2.2$, giving the combined mechanism shown in Fig. 4.8(f). The minimum weight becomes

$$G_{\min} = 4(\tfrac{7}{6}) + 3(\tfrac{1}{2}) = 6\cdot17$$

In any frame, it is possible to proceed step-by-step as shown, passing over the surface of the permissible region in such a way as always to decrease the total weight. In view of the uniqueness theorem, the minimum weight will ultimately be achieved by this procedure, although as a hand method it becomes laborious for all but the most simple frames. The method of inequalities may be applied to minimum weight design (Heyman,[3] Livesley,[4] and Heyman and Prager[5]) and leads to a more suitable procedure for solution by digital computer.

Effect of nonlinear weight functions and discrete sections

While the use of the linear weight function $g = kM$ appears at first sight to be unduly restrictive, it is to be noted that the more realistic relationship $g = kM^n$ (OAB in Fig. 4.10) may be represented quite closely over a wide range of plastic moment values M_1 to M_2 by an equation of the form

$$g = g_0 + kM$$

This leads, for the total weight of a structure, to an expression

$$G = G_0 + k \sum_j M_j l_i$$

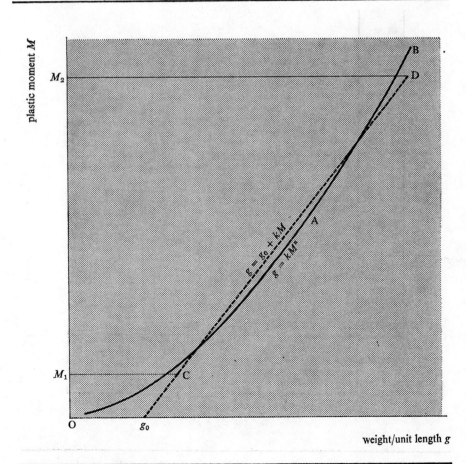

Figure 4.10 Linear approximation to plastic moment–weight relationship

The constant G_0 does not appear in the minimization process, and hence all the minimum weight theorems and procedures still apply.

Considering the problem more precisely, when a mechanism interaction diagram of the type shown in Figs. 4.3 and 4.6 is drawn with axes representing the weight of the members calculated on the assumption $g = kM^n$ (where $n < 1$) instead of $g = kM$, the mechanism lines that are not parallel to either axis become curved and concave to the origin as shown in Fig. 4.11. It is obvious that the uniqueness theorem no longer applies, since the four conditions of equilibrium, yield, mechanism, and plastic hinge may all be satisfied at more than one corner of the permissible region. Thus, if ab is the absolute minimum weight line with the minimum weight solution at point c, the parallel line $a'b'$ passes through c' and at both points c and c' the mechanisms may be combined

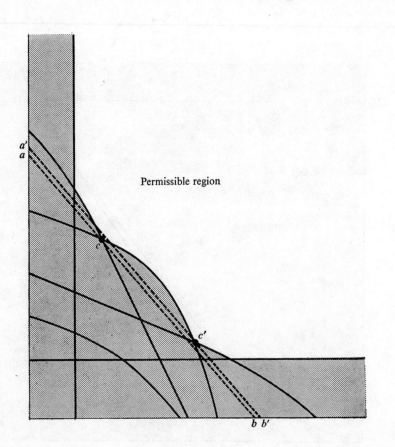

Permissible region

Figure 4.11 Minimum weight interaction diagram for nonlinear plastic moment–weight relationship

in such a way as to satisfy the plastic hinge condition. While the upper bound theorem still applies, the lower bound theorem does not. However, the local minima such as point c' appear unlikely to have a weight very much higher than the true minimum weight at c, and this is confirmed by calculations for specific cases.

The problem of minimum weight design when allowance is made for the availability of members in discrete sizes has been discussed by Toakley.[6] He describes a computer procedure for determining the absolute minimum weight in such cases, but finds that the demands on computer time and capacity are very much greater than when a continuous linear weight function is assumed. The best method in practice is first to solve any problem on the assumption that a continuous range of sections exists. If any particular member then has a moment requirement of M, with the next available sections below and above having plastic moment values M_L and M_U, the design tentatively adopted is M_U if $M > aM_L + (1 - a)M_U$, and M_L if $M < aM_L + (1 - a)M_U$ where a is an empirical constant. Practice will indicate a suitable value for a, but 0·3–0·4 will usually be appropriate. The structure so designed is then checked for failure load, and if this is less than the required value, one or more sections that have been given a plastic moment less than the linear analysis indicated are increased to the next section. The value chosen for a should be such that the large majority of structures are satisfactory at the first check.

Complete exploration of minimum weight frames for rectangular portal frames

Foulkes[1] has investigated the complete range of minimum weight, fixed-base rectangular portal frames loaded as shown in Fig. 4.12, and his complete results are shown in Fig. 4.13. This is an interesting example of the application of the minimum weight theorems, and the derivation of the limits

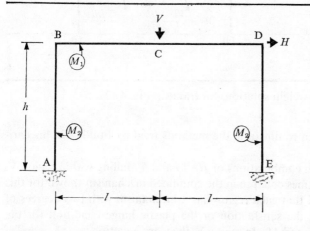

Figure 4.12 Portal frame with variable loading and height to span ratios

8—P.T.S.

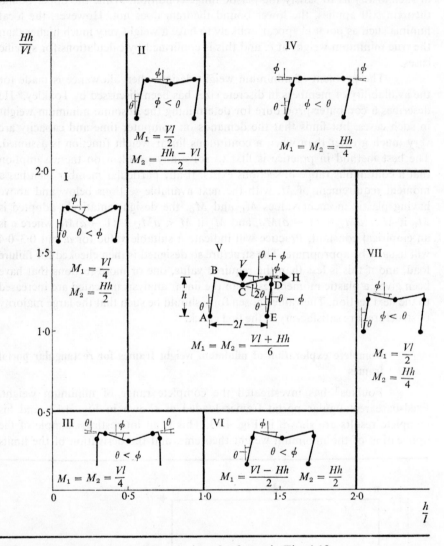

Figure 4.13 Minimum weight solutions for frame in Fig. 4.12

of region V will be given to illustrate the methods used by Foulkes in his derivation of the diagram.

For frames with combinations of Hh/Vl and h/l falling within region V, the minimum weight frames collapse in the combined mechanism shown for this region. It is to be noted that each region has a mechanism with two degrees of freedom, thus allowing the satisfaction of the plastic hinge condition for the two variable sections M_1 and M_2. In region V, the work equation is

$$Vl\theta + Hh\theta = 6M_1\theta = 6M_2\theta \tag{4.26}$$

irrespective of the value of ϕ, but to satisfy the plastic hinge condition, it is necessary to have

$$\frac{3\theta + \phi}{3\theta - \phi} = \frac{h}{l} \tag{4.27}$$

Since, if plastic hinges of the correct sign are to occur in both beam and column at D, $-\theta < \phi < \theta$,

$$\frac{1}{2} < \frac{h}{l} < 2 \tag{4.28}$$

The bending moment at B may be obtained by considering a virtual work equation, using either a beam or a sway mechanism. Using the beam mechanism,

$$M_{\mathrm{B}}\theta + 3M_1\theta = Vl\theta \tag{4.29}$$

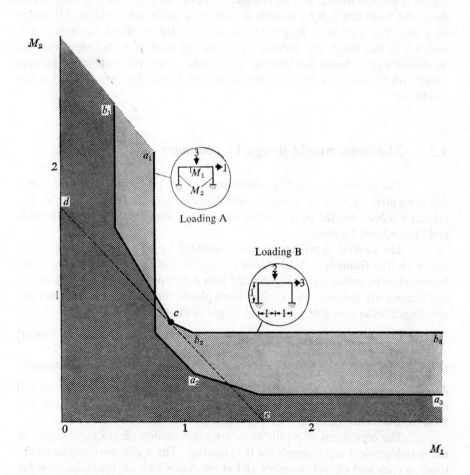

Figure 4.14 Minimum weight interaction diagram for frame with two sets of loadings

The condition $-M_1 < M_B < M_1$, applied to Eqns (4.26) and (4.29), leads to

$$\frac{1}{2} < \frac{Hh}{Vl} < 2 \tag{4.30}$$

thus establishing the limits of region V. The limits of the other regions are obtained similarly.

Minimum weight under alternating loads

When more than one combination of loads act on a structure, the minimum weight frame may be established by constructing the boundaries of the permissible regions for the separate sets of loads. Thus, suppose the portal frame in Fig. 4.5(a) may be subjected either to loading A or to loading B, as shown in Fig. 4.14. Loading A is the same as that for which the permissible region is derived in Fig. 4.6, reproduced as $a_1a_2a_3$ in Fig. 4.14. Loading B produces the boundary $b_1b_2b_2$, so that the resulting permissible region is bounded by a_1cb_3. The minimum weight line de touches the resultant boundary at c, which is in fact inside the permissible region for loading A and represents the minimum weight frame for loading B. In some cases, the resultant minimum weight frame may not be a minimum weight frame for any single loading condition.

4.5   Minimum weight design by computer

The minimum weight problem may be set up quite straightforwardly for computer solution by using standard algorithms that are available. The subject has been treated by a number of authors, including Livesley,[4] Heyman and Prager,[5] and Foulkes.[7]

The general approach may be illustrated by reference to the frame in Fig. 4.15. The frame is to be designed to support the given loads, the members (assumed to be prismatic) being divided into n groups. All the members in any one group i are required to have the same plastic moment M_{pi}, the weight per unit length being $g_i = kM_{pi}$. The total weight of the structure is then

$$G = k \sum_i \left(M_{pi} \sum_j L_{ij} \right) \tag{4.31}$$

where $\sum_j L_{ij}$ denotes the sum of the lengths of members with plastic moments M_{pi}. The design problem is to choose that set of plastic moments M_{pi} which will minimize the weight function G and enable the conditions of equilibrium and yield to be satisfied for the given set of loads.

The conditions of equilibrium are most readily set up by considering the m independent mechanisms for the structure. The p unknown moments are those at each end of each member and at the centre of each beam, less one for each pin joint. Thus, in Fig. 4.15,

$$p = (5 \times 3) + (8 \times 2) - 3 = 28$$

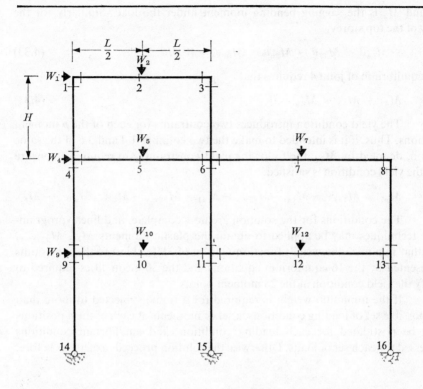

Figure 4.15 Multi-storey frame for minimum weight design

The degree of redundancy r is

$$r = (5 \times 3) - 3 = 12$$

so that the number of independent mechanisms is

$$m = p - r = 16$$

These are made up as follows:

> beam mechanisms 5
> sway mechanisms 3
> joint rotations 8.

There are 16 equations of equilibrium, obtained one from each independent mechanism. Thus, if the load W_2 descends a distance $L\theta/2$ due to the corresponding beam mechanism, the equation of equilibrium becomes

$$-M_{13}\theta + 2M_2\theta + M_{31}\theta = \frac{W_2 L}{2}\theta \qquad (4.32)$$

where M_{13}, M_{31} denote the clockwise moments acting on the ends of beam

1.3 and M_2 is the sagging bending moment under the load. Similarly for the sway of the top storey,

$$-M_{14}\theta - M_{41}\theta - M_{36}\theta - M_{63}\theta = W_1 H\theta \qquad (4.33)$$

The equilibrium of joint 4 requires that

$$M_{41} + M_{46} + M_{49} = 0 \qquad (4.34)$$

The yield condition introduces two contraints for each of the p moment positions. Thus, if it is intended to make the two columns 1.4 and 3.6 of the same section, denoted by M_{p7}, the following eight constraints are required to ensure that the yield condition is satisfied.

$$M_{p7} \geqslant M_{14}, \quad -M_{14}, \quad M_{41}, \quad -M_{41}, \quad M_{36}, \quad -M_{36}, \quad M_{63}, \quad -M_{63}.$$

The conditions for the solution are now complete, and linear programming techniques may be applied to obtain the plastic moments M_{p1}, M_{p2}, ..., M_{pn} that minimize the weight function G [Eqn (4.31)] subject to the constraints represented by the 16 equilibrium equations and the 56 inequalities required to satisfy the yield condition at the 28 moment points.

If the minimum weight is required for a frame subjected to more than one possible set of loading conditions, a set of moments at each of the p positions must be postulated for each loading condition, and equilibrium conditions expressed for each set of loads. Otherwise the solution proceeds exactly as before.

4.6   Minimum weight frames of continuously varying cross section

In frames composed of prismatic members, full plasticity is confined to discrete cross sections except for particular cases where a length of member is subjected to a uniform moment. If the possibility is admitted of allowing continuously varying members, every cross section of a minimum weight frame designed for a single set of loads must become fully plastic at the design load intensity, since otherwise reductions of cross sections would be possible. The discrete plastic hinge condition represented by Eqn 4.12 is therefore replaced by

$$d\phi = \alpha ds$$

where, in the collapse mechanism, $d\phi$ is the plastic rotation over a length ds. But $d\phi/ds = \kappa$, where κ is the curvature over the length ds. Hence, the collapse mechanism must be capable of assuming a form in which the curvature of *all* the members is *everywhere* constant. Since the curvature must have everywhere the same sign as the applied moment, the curvature changes sign along any member wherever the moment changes sign. It is to be noted that no finite rotations are allowed, since this would correspond to infinite curvature.

Consider as an example the propped cantilever AB in Fig. 4.16(a),

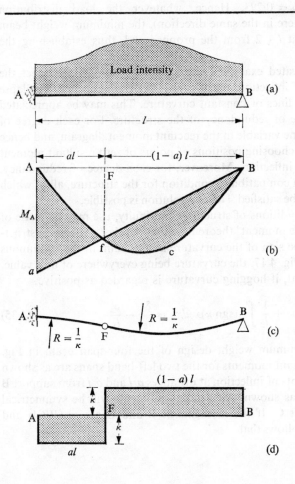

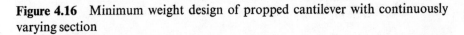

Figure 4.16 Minimum weight design of propped cantilever with continuously varying section

subjected to any varying distributed load as shown. Let AcB in Fig. 4.16(b) be the free moment diagram for this load, and suppose that, for the minimum weight design, the moment at A is M_A. The resultant applied moments (and consequently the design plastic moments of the beam) are then given by the shaded area. The deflected form of the beam at collapse must be as shown in Fig. 4.16(c), with a point of inflection at F where the required plastic moment is zero. Since the tangent at A intersects the beam centre line again at B, the moment of the curvature diagram [Fig. 4.16(d)] about B must be zero. (This is an adaptation of the 'moment-area' method of analysis for elastic structures.) Hence

$$al\kappa \left(1 - \frac{a}{2}\right)l = (1 - a)\, l\kappa \left(\frac{1 - a}{2}\right)l$$

whence $a = 1 - (1/\sqrt{2}) = 0\cdot293$. Hence, whatever the load distribution (provided it acts everywhere in the same direction), the minimum weight beam will have zero moment at $l/\sqrt{2}$ from the propped end, thus establishing the design.

In more complicated examples, the load distribution does affect the positions of the points of inflection, but it is always possible to find a collapse mechanism composed of lines of constant curvature. This may be appreciated by considering the degree of redundancy of the structure. For each degree of redundancy there exists one variable in the reactant moment diagram, and hence one degree of freedom in choosing positions of points of zero resultant moment which are also points of inflection. Moreover, for each degree of redundancy, there exists a deformation compatibility condition for the structure, all of which conditions may therefore be satisfied. Hence a solution is possible.

In solving the conditions of structural continuity, use may be made of an adaptation of the 'three moment' theorem used in the elastic theory of structures. If (sgn κ) denotes the sign of the curvature at any point in the continuous two-span beam ABC in Fig. 4.17, the curvature being everywhere of unit value, then it is readily shown that, if hogging curvature is regarded as positive,

$$\frac{1}{l_1} \int_0^{l_1} (\text{sgn } \kappa) \, x_1 dx_1 + \frac{1}{l_2} \int_0^{l_2} (\text{sgn } \kappa) x_2 dx_2 = \frac{\delta_2}{l_2} - \frac{\delta_1}{l_1} \qquad (4.35)$$

Consider the minimum weight design of the four-span beam in Fig. 4.18(a). The free and reactant moments for the two left-hand spans are as shown in Fig. 4.18(b), giving points of inflection at distances a_1l and a_2l from support B and a_3l from support C as shown. All bending moments will be symmetrical about the central support C. If the moments at B and C are $m_1(Wl/4)$ and $m_2(Wl/4)$ respectively, it follows that

$$a_1 = \frac{m_1}{2 + m_1}$$

$$a_2 = \frac{m_1}{2 + m_1 - m_2}$$

$$a_3 = \frac{m_2}{2 - m_1 + m_2}$$

The curvatures are as shown in Fig. 4.18(c), whence Eqn (4.35) applied successively to spans ABC, BCD gives

$$\left\{ \frac{(1 - a_1)^2}{2} - a_1 \left(1 - \frac{a_1}{2} \right) \right\}$$

$$+ \left\{ -a_2 \left(1 - \frac{a_2}{2} \right) + (1 - a_2 - a_3) \left(\frac{1 - a_2}{2} + \frac{a_3}{2} \right) - \frac{a_3^2}{2} \right\} = 0$$

$$- \frac{a_2^2}{2} + (1 - a_2 - a_3) \left(\frac{a_2}{2} + \frac{1 - a_3}{2} \right) - a_3 \left(1 - \frac{a_3}{2} \right) = 0$$

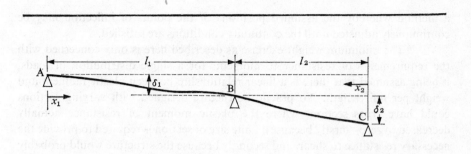

Figure 4.17 General deformation state for plastic minimum weight design with continuously varying section

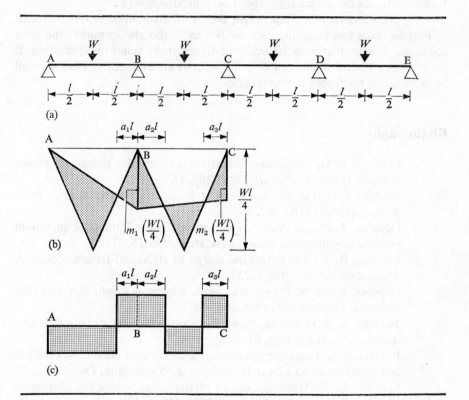

Figure 4.18 Minimum weight design of four-span beam of continuously varying section

Five equations are available for the determination of the five unknowns a_1, a_2, a_3, m_1, and m_2. Solution by trial and error gives $m_1 = 0.525$, $m_2 = 0.443$.

Obviously, the numerical labour involved in obtaining solutions increases rapidly with the complexity of the structure. Heyman[9] has suggested a method whereby the assumed positions of the points of inflection may be continuously adjusted until the continuity conditions are satisfied.

The minimum weight exercise as described here is only concerned with the requirements of resistance to moments for a single distribution of loads, it being assumed that there is a linear relationship between plastic moment and weight per unit length. No practical structure designed with varying sections could have cross sections where the plastic moment of resistance actually decreased to zero—firstly because a finite size of section is required to provide the necessary resistance to shear, and secondly because the structure would probably have inadequate carrying capacity for any distribution of loads other than that for which the structure was designed. Thus, for example, the simply supported four-span beam previously considered [Fig. 4.18(a)] would actually be in a state of unstable equilibrium as designed, and would not in general support any loading that was not symmetrical about the central support C.

The absolute minimum weight design of fixed-ended beams subjected to moving loads has been discussed by Horne,[10] who also considers the most economic designs that may be achieved under static (uniformly distributed) loads when a limited number of differing sections are allowed, as when a rolled beam section is reinforced by flange plates.

Bibliography

1   Foulkes, J. D. Minimum weight design and the theory of plastic collapse. *Quart. Appl. Math.*, 1953, **10**, 347.

2   Foulkes, J. D. The minimum weight design of structural frames. *Proc. R. Soc.*, 1954, **223** [A], 482.

3   Heyman, J. Plastic design of beams and plane frames for minimum material consumption. *Quart. Appl. Math.*, 1951, **8**, 373.

4   Livesley, R. K. The automatic design of structural frames. *Quart. J. Mech. Appl. Math.*, 1956, **9**, 257.

5   Heyman, J. and W. Prager. Automatic minimum-weight design of steel frames. *J. Franklin Inst.*, 1958, **266**, 339.

6   Toakley, A. R. Optimum design using available sections. *J. struct. Div. Am. Soc. civ. Engrs*, 1968, **94** [ST5], 1219.

7   Foulkes, J. D. Linear programming and structural design. Proceedings 2nd Symposium on Linear Programming, Washington, DC, 1955, 177.

8   Livesley, R. K. Optimum design of structural frames for alternative systems of loading. *Civ. Engng*, 1959, **54**, 737.

9   Heyman, J. On the absolute minimum weight design of framed structures. *Quart. J. Mech. Appl. Math.*, 1959, **12**, 314.

10   Horne, M. R. Determination of the shape of fixed-ended beams for maximum economy according to the plastic theory. Preliminary Publication of the 4th Congress of the International Association of Bridge and Structural Engineering, Cambridge, 1952, 111. Also Final Report, 1952, 119.

Problems

4.1 A continuous beam ABC, fixed in position and direction at A and simply supported at B and C, has spans AB = $3L$ and BC = $4L$. Loads of magnitude $2W$ are applied at each of the third points in span AB (that is, at distances of L from A and B) and a load of magnitude W is applied at mid-span in BC. If the spans AB and BC have plastic moments of M_1 and M_2 respectively, find M_1 and M_2 for minimum weight consumption, assuming a linear relationship between plastic moment and weight per unit length.

4.2 A continuous beam ABCD has spans AB = CD = L_1 and BC = L_2, is simply supported at A, B, C, and D, and supports concentrated loads W_1 at the centres of spans AB and CD and W_2 at the centre of span BC. Find the minimum weight designs in terms of the plastic moments M_1 for spans AB and CD, and M_2 for span BC, for all values of L_1/L_2 and all positive values of W_1L_1/W_2L_2.

4.3 The fixed-base, two-storey symmetrical frame in Fig. E4.1 is to be designed for minimum weight for the loads shown, assuming members of uniform section with plastic moments M_1, M_2, M_3, and M_4 as shown. Find the values of M_1 to M_4, assuming a linear relationship between plastic moment and weight per unit length.

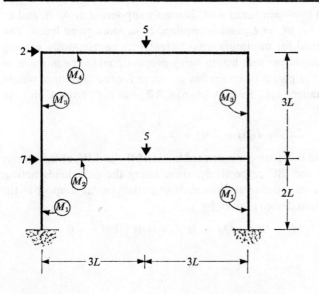

Figure E4.1

4.4   A rectangular fixed-base frame has a span of $4L$ and a column height of $3L$ with plastic moments of M_1 in the columns and M_2 in the beam. It is subjected to the two following alternative loading systems.

Loading A Central vertical beam load $8W$, horizontal load at beam level $5W$.

Loading B Central vertical beam load $11W$, no horizontal load.

Draw on a graph of M_1 versus M_2 the combinations of M_1 and M_2 for various mechanisms under the two loading combinations. Hence define the permissible region and, assuming a linear relationship between plastic moment and weight per unit length, determine the minimum weight design.

4.5   A pin-based rectangular portal frame ABCD, consisting of columns AB, CD of height H and a beam BC of span L, is designed for minimum total weight for a symmetrical vertical loading distribution on the beam. The section varies continuously, the weight per unit length being $g_0 + g_1 M_p$, where M_p is the plastic moment of resistance. Show that, provided $L > H$, points of contraflexure occur in the beam BC at distances of $(L - H)/4$ from B and C.

If the beam carries a uniformly distributed load of w per unit length, show that, provided $L \geqslant H$, the minimum weight is given by

$$(L + 2H) g_0 + (3L^3 + 9L^2H - 3LH^2 - H^3) wg_1/96$$

and that when $L \leqslant H$, the minimum weight is

$$(L + 2H) g_0 + L^3 wg_1/12$$

4.6   A continuous two-span beam ABC is simply supported at A, B, and C, where $AB = L_1$, $BC = L_2$ and is required to support given loads. The beam is designed for minimum total weight with continuously varying section, the weight per unit length being proportional to the moment of resistance. If the points of contraflexure occur for the minimum weight design at distances $a_1 L_1$ from A in span AB, and $a_2 L_2$ from C in span BC, show that

$$L_1(l - 2a_1^2) + L_2(l - 2a_2^2) = 0$$

If $L_1 = L_2$ and there are concentrated loads of W_1 and W_2 at the centres of spans AB and BC respectively, these being the only loads acting, show that the moment of resistance M over the central support in the minimum weight design is given by

$$16M^2 (M + W_1L_1) (M + W_2L_2) - W_1^2 W_2^2 L^4 = 0$$

Chapter 5
Variable repeated loading

5.1   Introduction

When a rigid frame is loaded beyond the elastic limit, the bending moments in at least part of the structure will lie between the *moment at first yield M_y* and the *plastic moment M_p* (Fig. 5.1). When the structure is unloaded, the changes of

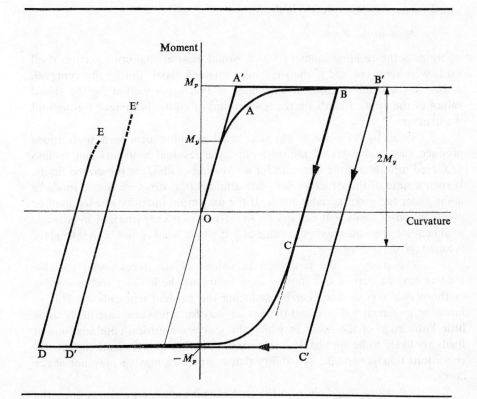

Figure 5.1   Moment–curvature relations under reversed bending

bending moment will certainly at first take place elastically, and these changes will not usually be in proportion to the bending moments existing at the end of the loading process unless the structure is a statically determinate one. Since the total change of bending moment that may take place elastically at any cross section is $2M_y$ (usually considerably greater than M_p), it is found that most structures, when unloaded under proportional loading, do so elastically, leaving a set of residual moments in equilibrium with zero external loads. Thus the fixed-ended beam AB in Fig. 5.2(a), of length L, loaded to collapse under a uniformly distributed load of magnitude $W = 16M_p/L$ [giving the bending moment diagram in Fig. 5.2(b)] will, when unloaded, undergo changes of moment at the ends of $WL/12 = \frac{4}{3}M_p$ (sagging), and at the centre of $WL/24 = \frac{2}{3}M_p$ (hogging), according to the bending moment diagram in Fig. 5.2(c). The residual moment state is then a uniform sagging moment of $\frac{1}{3}M_p$, as shown in Fig. 5.2(d). On reloading up to the limit of the failure load, changes of moment take place elastically.

It is evident that, if a structure is subjected to a succession of differing load distributions varying between prescribed limits, it is not so easy to discover whether a state can be reached such that all subsequent load applications produce only elastic changes of stress. If the structure does reach such a state, then the total bending moment M at any cross section when the external loads have stated values may be expressed in the form

$$M = m + \mathcal{M}$$

where m is the residual moment which would exist at that cross section if all loads were removed and if the structure remained elastic during this removal, and $\mathcal{M}$ is the bending moment calculated at that cross section for the stated values of the applied loads on the assumption of elastic behaviour throughout the structure.

Since, by definition of the state reached, subsequent load applications produce elastic changes of moment only, the residual moments can remain unaltered throughout the structure for all loadings within the prescribed limits. If such a state of the structure is in fact attained, the structure is said to *shake down* under the given variable loads. If the maximum intensity of each load or combination of loads that can act on the structure is characterized by a single load factor λ, then the maximum value of λ at which 'shakedown' can take place is called the *shakedown limit*.

The study of the post-yield behaviour of structures under variable loading may be carried out step by step, following the loading and unloading of the plastic regions and thereby deducing the residual moments m. This is, however, extremely tedious, and there is anyway for a practical case likely to be little knowledge of the order in which the successive critical combinations of loads are likely to be applied. It is important therefore to investigate the general conditions that govern the possibility that a structure may or may not shake down.

One controlling factor on the shakedown limit is an obvious one—the incidence of *alternating yield*. If $\mathcal{M}^{\max}$ and $\mathcal{M}^{\min}$ denote the (algebraic) maximum

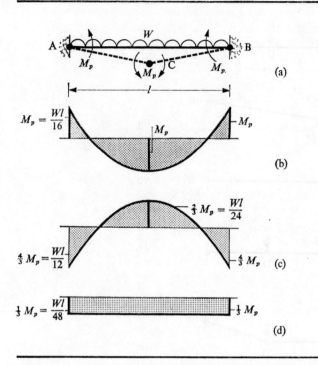

Figure 5.2 Residual moments in beam under uniformly distributed load

and minimum elastically calculated moments that can be induced at a given cross section under the prescribed variable loads, then it would certainly be impossible for shakedown to occur if the difference between these moments exceeded $2M_y$ (Fig. 5.1). Hence one necessary condition for shakedown is that, at every cross section,

$$\mathcal{M}^{\max} - \mathcal{M}^{\min} \leqslant 2M_y \qquad (5.1)$$

If this condition is violated, yield will occur in the extreme fibres at the critical cross sections, alternately in tension and compression as the loads are varied. This is obviously a particular case of the fatigue problem, namely the case when the stress limits are $\pm \sigma_y$. Because plastic deformation occurs alternately in tension and compression, the fatigue life will depend on the degree of strain as well as stress reversal (Royles[1]), and work-hardening will lead to changes in the stress–strain behaviour, away from the simple elastic–pure plastic relation. Ultimately, failure of the structure will take place by the breakdown of the material, but usually only after a large number of loading cycles.[1] This form of failure is called *alternating yield*.

There is another form of failure due to variable loading that is not so obvious as alternating yield. If the fixed-ended beam AB in Fig. 5.3(a), having a uniform plastic moment M_p, is subjected alternately to loads W at the third points C and D, static plastic collapse will occur for the load in either position

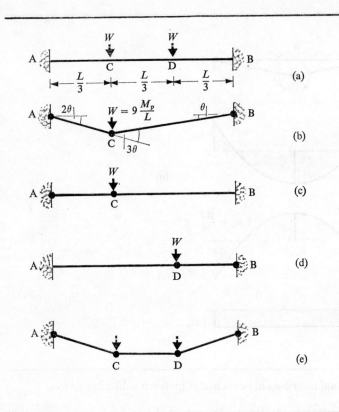

Figure 5.3 Incremental collapse of fixed-ended beam loaded alternately at third points

when $W = 9 \cdot 00 M_p/L$ [Fig. 5.3(b)]. In an initially stress-free beam, yield first occurs at A with the load at C, or at B with the load at D, at a load $W = 6 \cdot 75$ M_y/L, where M_y is the yield moment. When W has a value lying between $6 \cdot 75$ M_y/L and $8 \cdot 10 M_p/L$, the beam will shake down under the alternating loads, but for W lying between $8 \cdot 10 M_p/L$ and $9 \cdot 00 M_p/L$, shakedown cannot take place, despite the non-occurrence of alternating yield. This is because of the onset of *incremental collapse*. With W at C [Fig. 5.3(c)], limited plastic rotation occurs at A and C, but the plastic hinge rotations are *contained* (that is, prevented from increasing indefinitely) because the other fixed end B does not become plastic. The subsequent application of the load W at D [Fig. 5.3(d)] leads to contained plastic hinge rotations at D and B. This alters the residual stress state left after the removal of W from C, so that the reapplication of W at C is accompanied by further plastic deformations at A and C. The resultant effect is a progressive build-up of permanent deflections as though the beam were deforming in the plastic collapse mechanism shown in Fig. 5.3(e). This is an *incremental collapse mechanism* which operates if the load W exceeds the shakedown limit, which is in this case $8 \cdot 10 M_p/L$. The more closely the load approaches the static collapse value $9 \cdot 00 M_p/L$, the greater the increase of permanent deformation per loading cycle.

To illustrate the incremental collapse phenomenon more clearly, a simple example of a two-span beam is now discussed.

5.2 Behaviour of two-span beam

The beam ACE in Fig. 5.4(a) is of uniform section with plastic moment M_p and is subjected to central loads in the two spans of $\lambda_1 M_p/L$ and $\lambda_2 M_p/L$ as shown. The load $\lambda_2 M_p/L$ on the right-hand span is first applied, and the load $\lambda_1 M_p/L$ on the left-hand span then alternates in the sequence $\lambda_1 = 0 \rightarrow \lambda_2 \rightarrow 0 \rightarrow \lambda_2$ etc. It is assumed that the supports prevent upward as well as downward movement of the beam. Static collapse, either of the right-hand span or of both spans together, occurs when $\lambda_1 = \lambda_2 = 6 \cdot 00$, as shown in Fig. 5.4(b).

The elastic bending moments due to loads acting separately on the two spans are shown in Fig. 5.4(c) and (d). The maximum moment when $\lambda_1 = 0$, $\lambda_2 = \lambda$ occurs under the load at D, and is of magnitude $0 \cdot 203 \lambda M_p$, while with both loads acting ($\lambda_1 = \lambda_2 = \lambda$), the maximum moment occurs over the support at C and is of magnitude $2 \times 0 \cdot 094 \lambda M_p = 0 \cdot 188 \lambda M_p$. Hence, neglecting the spread of plasticity away from sections of maximum moment, that is, assuming $M_y = M_p$, the first plastic hinge that can form as λ is gradually increased occurs at D with loading $\lambda_1 = 0$, $\lambda_2 = 4 \cdot 93$, the deflection at D and the bending moments then being as given in row 1 of Table 5.1. If the load factor λ_2 is then increased by some quantity λ' while λ_1 remains zero, the *additional* moments and deflections will develop without any change of moment at D. These additional moments and deflections may therefore be calculated by analysing an elastic beam ACE [Fig. 5.4(e)], having a frictionless hinge at D. The state of the beam when $\lambda_1 = 0$, $\lambda_2 = 4 \cdot 93 + \lambda'$ is then as summarized in row 2 of Table 5.1.

We now add load to the left-hand span, keeping the load constant on the right-hand span. Since the plastic hinge at D must now unload, the beam behaves elastically. The additional moments and deflections may therefore be derived from the elastic solution in Fig. 5.4(d), giving, when the loads on the two spans have become equal, the resultant moments shown in row 3 of Table 5.1. The greatest moment now occurs over the central support at C, and this becomes just equal to the plastic value at full loading ($\lambda_1 = \lambda_2 = 4 \cdot 93 + \lambda'$) when

$$0 \cdot 922 + 0 \cdot 594 \lambda' = 1 \cdot 000$$

that is, when $\lambda' = 0 \cdot 13$ or $\lambda_1 = \lambda_2 = 5 \cdot 06$. Provided this load intensity is not exceeded, the subsequent removal and reapplication of the load on the left-hand side merely results in elastic changes of bending moment and deflection according to the pattern in Fig. 5.4(d). Hence the deflection at D and the bending

9—P.T.S.

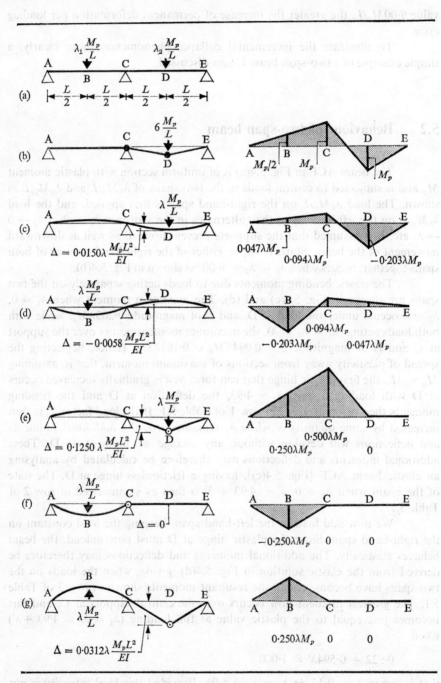

Figure 5.4 Incremental collapse of two-span simply supported beam ($\lambda_2 =$ constant, $\lambda_1 = 0 \rightarrow \lambda_2 \rightarrow 0 \rightarrow \lambda_2$, etc.)

moments oscillate between the values given in rows 2 and 3 of Table 5.1. The structure has, in fact, shaken down. If the loads are removed, all unloading occurs elastically and the residual moments and deflexion at D are as shown in row 4 of Table 5.1. Note that by adding the unloading moments to the reactant bending moments and deflexion the residual state moments and deflexion are given in rows 5 to 7.

We now consider what happens when λ_2 has a value intermediate between 0·06 and the static collapse load factor 0·06. Table 5.2 shows the results for $\lambda_2 = 5\cdot30$. Row 1 gives the situation at the first application of the load on the right-hand side; λ_2 is increased to a maximum of $\lambda = 5\cdot30$ by putting $\lambda = 5\cdot30 = 4\cdot93 + 0\cdot57$. Initially the left-hand span progresses plastically according to the changes in fig. 5.4(b), but, by considering the moment at C, it is seen that this becomes plastic at a value of the load factor λ given by

$$0\cdot750 - 0\cdot044\lambda = 1\cdot000$$

that is

Thereafter during the loading of the right-hand span the moment at C remains constant at the plastic value, so that the degree of moment take place as though the left-hand span were simply supported, as shown in Fig. 5.4(b). The resultant moments throughout the beam and the deflexion at D when $\lambda = \lambda_2 = 5\cdot30$ are as shown in row 2 of Table 5.2.

The subsequent behaviour depends on whether the right-hand span first closes. changes according to the changes in Fig. 5.4(b), for when λ_2 has been reduced from 5·30 to 4·66, the moment at D in the right-hand span becomes plastic, and the subsequent unloading of the left-hand span causes changes according to the pattern in Fig. 5.4(d). When all the loads on the right-hand span has been removed, the resulting magnitudes are as shown in row 3 of Table 5.2. Row 4 represents exactly the same situation as row 5 as far as moments are concerned, but the interesting phenomenon has arisen that the central deflexion in the right-hand span has increased by 0·044 $(V_p L^2/EI)$. It is evident that further cycles of loading and unloading of the left-hand span must each produce similar increases of

Table 5.1 Behaviour of two-span beam in Fig 5.4 when $0 < \lambda_2 \leqslant 5\cdot06$

	Operation	λ_1	λ_2	$\dfrac{M_B}{M_p}$	$\dfrac{M_C}{M_p}$	$\dfrac{M_D}{M_p}$	$\left(\dfrac{EI}{M_pL^2}\right)\Delta$
(1)	4·93 (c)	0	4·93	0·231	0·461	−1·000	0·074
(2)	(1) + λ' (e)	0	4·93 + λ'	0·231 + 0·250λ'	0·461 + 0·500λ'	−1·000	0·074 + 0·125λ'
(3)	(2) + (4·93 + λ') (d)	4·93 + λ'	4·93 + λ'	0·769 + 0·047λ'	0·922 + 0·594λ'	−0·769 + 0·045λ'	0·074 + 0·119λ'
(4)	(2) − (4·93 + λ') (c)	0	−(4·93 + λ')	0·203λ'	0·406λ'	0·203λ'	0·110λ'
(5)	(2), λ' = 0·13	0	5·06	0·263	0·526	−1·000	0·090
(6)	(3), λ' = 0·13	5·06	5·06	0·775	1·000	−0·763	0·061
(7)	(4), λ' = 0·13	0	0	0·026	0·053	0·026	0·014

moments oscillate between the values given in rows 2 and 3 of Table 5.1. The structure has, in fact, shaken down. If both loads are removed, all unloading occurs elastically, and the residual moments and deflection at D are as shown in row 4 of Table 5.1. When λ_2 has its limiting value of 5·06, the variation of bending moments and deflection and the residual state are as given in rows 5 to 7.

We now consider what happens when λ_2 has a value intermediate between 5·06 and the static collapse load factor 6.00. Table 5.2 shows the results for $\lambda_2 = 5.50$. Row 1 gives the situation at the first application of the load on the right-hand side, and is derived from row 2 of Table 5.1 by putting $\lambda' = 5\cdot50 -4\cdot93 = 0\cdot57$. Initially the addition of load to the left-hand span produces changes according to the elastic solution in Fig. 5.4(d), but, by considering the moment at C, it is seen that this becomes plastic at a value of the load factor λ_1 given by

$$0\cdot750 + 0\cdot094\lambda_1 = 1\cdot000$$

that is, $\lambda_1 = 2\cdot66$

Thereafter, during the loading of the left-hand span, the moment at C remains constant at the plastic value, so that changes of moment take place as though the left-hand span were simply supported, as shown in Fig. 5.4(f). The resultant moments throughout the beam and the deflection at D when $\lambda_1 = \lambda_2 = 5\cdot50$ are as shown in row 2 of Table 5.2.

The subsequent removal of the load on the left-hand span first causes changes according to the elastic solution of Fig. 5.4(d), but when λ_1 has been reduced from 5.50 to 2.66, the moment at D in the right-hand span becomes plastic, and the subsequent unloading of the left-hand span causes changes according to the pattern in Fig. 5.4(g). When all the load on the left-hand span has been removed, the resulting condition of the beam is as shown in row 3 of Table 5.2. Row 3 represents exactly the same solution as row 1 as far as moments are concerned, but the interesting phenomenon has arisen that the central deflection in the right hand span has increased by 0·089 (M_pL^2/EI). It is evident that further cycles of loading and unloading of the left-hand span must lead to similar increases of

Table 5.2 Behaviour of two-span beam in Fig. 5.4 when $\lambda_2 = 5\cdot50$

	Operation	λ_1	λ_2	$\dfrac{M_B}{M_p}$	$\dfrac{M_C}{M_p}$	$\dfrac{M_D}{M_p}$	$\left(\dfrac{EI}{M_pL^2}\right)\Delta$
(1)	4·93 (c) +0·57 (e)	0	5·50	0·375	0·750	−1·000	0·146
(2)	(1) +2·66 (d) +2·84 (f)	5·50	5·50	−0·875	1·000	−0·875	0·130
(3)	(2) −2·66 (d) +2·84 (g)	0	5·50	0·375	0·750	−1·000	0·235
(4)	(3) +2·66 (d) +2·84 (f)	5·50	5·50	−0·875	1·000	−0·875	0·219
(5)	(4) −2·66 (d) +2·84 (g)	0	5·50	0·375	0·750	−1·000	0·324

deflection, as shown for the next cycle in rows 4 and 5. This is because of the irreversible plastic deformation that occurs at C during the later stages of the *loading* of the left-hand span, and at D during the later stages of the *unloading* of the left-hand span.

This pattern of incremental deformation will obviously occur at any value of λ_2 in excess of 5·06, which is therefore the shakedown limit. Inspection also shows that, in this example, the magnitude of the incremental deformation per cycle is directly proportional to the amount by which the load factor λ_2 exceeds the shakedown limit. This simple relationship arises because no further plastic hinges can form before the static collapse load is reached, and is not the case for all structures.

The shakedown limit has been established for a given order of loading for a given structure by following a step-by-step procedure. The question arises as to whether the same shakedown limit would apply for other orders of loading, and, if so, how the shakedown limit may be derived by a more convenient process. These matters may be resolved by considering the general theory of shakedown for rigid frames, to which we now turn our attention.

5.3   Theorems of shakedown

It may be shown that a rigid frame structure will approach a shakedown state under a set of variable repeated loads provided a set of residual moments m^* may be postulated such that, at every cross section,

$$\left.\begin{array}{l} m^* + \mathcal{M}^{\max} \leqslant M_p \\ m^* + \mathcal{M}^{\min} \geqslant -M_p \end{array}\right\} \tag{5.2}$$

$$\mathcal{M}^{\max} - \mathcal{M}^{\min} \leqslant 2M_y \tag{5.3}$$

where M_p is the plastic moment and $2M_y$ the range of moment over which an elastic moment–curvature relationship exists (Fig. 5.1). The moments $\mathcal{M}^{\max}$ and $\mathcal{M}^{\min}$ are respectively the maximum and minimum values calculated from elastic theory for an initially stress-free structure due to all the allowed combinations of applied loads.

A shakedown theorem for simple bar structures [involving inequalities corresponding to Eqn (5.2) only] was first given by Bleich,[2] and given more general proof by Melan.[3] Neal[4] first extended the theorem to rigid frame structures for members having the simplified moment–curvature relationship OA'B'C'D'E' in Fig. 5.1, in which case $2M_y = 2M_p$ and the inequality (5.3) is contained within the inequalities (5.2). Later, Neal[5] extended his treatment to deal with members having the more realistic moment–curvature relationship OABCDE in Fig. 5.1.

The following points about the shakedown theorem may be noted.

(a) The postulated residual moments m^* need not necessarily be the same as the residual moments m which would actually exist in the structure after it had shaken down. It is sufficient to be able to postulate any set of moments m^* satisfying inequalities (5.2) and (5.3).

(b) The shakedown condition is unaffected by initial stresses however caused, so that these have no effect on whether or not a structure can shake down under a given set of loads. Initial stresses may, however, affect the number of load variations which have to take place before a condition of shakedown is actually reached.

(c) The order in which loads are applied has no effect on whether a structure can shake down, although again the order of loading may influence the rapidity with which a shakedown state is reached.

(d) No statement is made in the theory that a condition of shakedown can in fact be reached in a finite number of load variations—only that a state of shakedown is approached. Nor is it stated whether or not the total plastic deformations are restricted to finite values when inequalities (5.2) and (5.3) can be satisfied. Neal[6] has, however, shown that in rigid frame structures the permanent deformations are in fact finite.

(e) The elastic moments $\mathcal{M}^{\max}$ and $\mathcal{M}^{\min}$ may include changes of moment induced by variations of temperature, in which case the term 'variable repeated loads' is extended to cover also 'variable repeated temperature states'. No further mention of this extension is made in the proofs that follow, but all statements will remain valid, except that when dealing with alternating yield, it is necessary to consider in detail the effect of temperature stresses within the cross section on the range of moment (corresponding to M_y) over which the moment–curvature relationship remains linear.

The shakedown theorem will be proved for the simplified moment–curvature relationship OA'B'C'D'E' in Fig. 5.1. The theorem may then be stated as follows.

Shakedown theorem

A rigid frame structure will approach a shakedown state under a set of loads varying between prescribed limits provided a set of residual moments m^* may be postulated such that, at every cross section, inequalities (5.2) are satisfied, where $\mathcal{M}^{\max}$ and $\mathcal{M}^{\min}$ are the maximum and minimum moments due to the given loads.

Proof. Let the residual moment actually existing at any section at any particular instant in the loading sequence be m. Then, if M is the total bending moment at that section and $\mathcal{M}$ is the elastic moment at the same section due to the current loading,

$$m = M - \mathcal{M} \tag{5.4}$$

It is to be noted that the moments m are in equilibrium with zero external loads.

Consider the necessarily finite positive quantity

$$U = \int \frac{(m - m^*)^2}{2EI} \, ds \tag{5.5}$$

where EI is the flexural rigidity of a member and s is distance measured along the member, the integration being performed over the whole structure. Suppose that, owing to any small change in loading, the moments M, $\mathcal{M}$, and m change by δM, $\delta \mathcal{M}$, and δm respectively. Then the change in U becomes

$$\delta U = \int \frac{(m - m^*)\delta m}{EI} \, ds \tag{5.6}$$

It will now be shown that δU is either zero or negative.

Any change δm in the residual moments m must arise as a result of increments of plastic deformation at one or more of those positions at which the plastic moment $\pm M_p$ happens to be reached under the given current loading for the current set of residual moments. Suppose these incremental plastic hinge rotations, occurring at positions h_j, are denoted by $\delta \theta_j$, which are therefore compatible with changes of curvature $\delta M/EI$. Since the changes of curvature $\delta \mathcal{M}/EI$ correspond to zero plastic hinge rotations (all $\mathcal{M}$ being calculated on the assumption of a completely elastic structure), curvatures $\delta(M - \mathcal{M})/EI = \delta m/EI$ are also compatible with incremental hinge rotations $\delta \theta_j$. Since the moments $(m - m^*)$ are in equilibrium with zero external loads, we may therefore write the virtual work equation

$$\int \frac{(m - m^*)\delta m}{EI} \, ds + \sum_j (m_j - m_j^*)\delta \theta_j = 0 \tag{5.7}$$

where m_j and m_j^* denote the values of m and m^* respectively at the positions h_j at which incremental plastic hinge rotations $\delta \theta_j$ have taken place. Hence from Eqns (5.6) and (5.7),

$$\delta U = -\sum_j (m_j - m_j^*)\delta \theta_j \tag{5.8}$$

Suppose that, at a particular section,

$$m_j < m_j^* \tag{5.9}$$

Then, from the first inequality of (5.2),

$$m_j + \mathcal{M}_j^{\max} < M_p \tag{5.10}$$

The maximum bending moment that can occur at the section in question for the current value m_j of the residual moment is, from Eqn (5.4), $M_j^{\max}$, where

$$m_j + \mathcal{M}_j^{\max} = M_j^{\max} \tag{5.11}$$

From Eqns (5.10) and (5.11),

$$M_j^{\max} < M_p \tag{5.12}$$

Since the maximum moment that can occur at the section under discussion is thus *less* than the plastic moment, the incremental plastic hinge rotation $\delta\theta_j$ can only arise with the minimum moment equal to $-M_p$, thus showing that $\delta\theta_j$ must be negative. Hence from Eqn (5.9),

$$(m_j - m_j^*)\delta\theta_j > 0 \tag{5.13}$$

Similarly, it may be shown that, when $m_j > m_j^*$, $\delta\theta_j$ is positive, whence Eqn (5.13) is still true. Obviously, when $m_j = m_j^*$,

$$(m_j - m_j^*)\delta\theta_j = 0 \tag{5.14}$$

Since either Eqn (5.13) or (5.14) is true for each position h_j, it follows from Eqn (5.8) that

$$\delta U \leqslant 0 \tag{5.15}$$

It is thus proved that δU is either zero or negative. If the structure behaves elastically so that there are no hinge rotations (all $\delta\theta_j$ zero), it follows from Eqn (5.8) that δU is zero and U remains constant. Hence U decreases whenever plastic deformation occurs, and remains constant when the structure behaves elastically. Since U is always positive finite, the structure must approach a state in which U either becomes zero or reaches a constant positive value. In either case, the structure approaches a state of shakedown. If U becomes zero, the actual residual moments m are equal everywhere to the postulated moments m^*, while if U is finite, the actual residual state differs from that postulated.

When the moment–curvature relation is of the form OABCDE (Fig. 5.1) with an elastic range $2M_y < 2M_p$, the structure will not shake down if the elastic range of moment at any section exceeds $2M_y$, whence inequality (5.3) must also be imposed to ensure the possibility of shakedown. In any particular case, the limitation on the shakedown load will be imposed by the onset of incremental collapse if any increase of load factor prevents inequalities (5.2) being satisfied, while still allowing the satisfaction of inequality (5.3). Conversely, if an increase of load factor causes the violation of (5.3) while still allowing (5.2) to be satisfied, failure is by alternating yield.

Upper and lower bounds on shakedown limits

The intensity of variable repeated loading on a given structure may be defined by a common load factor λ, so that $\lambda\mathcal{M}^{\max}$ and $\lambda\mathcal{M}^{\min}$ represent the maximum and minimum elastic moments for the load factor λ. The problem is then to derive upper and lower bounds on λ_s, the shakedown limit, where λ_s is the maximum value of λ for which shakedown can occur.

Lower bound

A lower bound on the shakedown limit is given by any load factor λ

for which a set of residual moments $m*$ satisfying the inequalities

$$\left. \begin{array}{l} m* + \lambda \mathcal{M}^{\max} \leqslant M_p \\ m* + \lambda \mathcal{M}^{\min} \geqslant -M_p \end{array} \right\} \tag{5.16}$$

$$\lambda(\mathcal{M}^{\max} - \mathcal{M}^{\min}) \leqslant 2M_y \tag{5.17}$$

can be found.

One lower bound is obviously obtained by putting $m* = 0$. The load factor λ is then such that the maximum range of moment does not exceed $2M_y$, while the maximum moment (assuming zero initial stress) as calculated from elastic theory does not exceed M_p. This is a useful lower bound, and may in some design situations eliminate the need for more complete analyses.

Upper bounds

An upper bound on the shakedown load can be obtained by considering either alternating yield or incremental collapse. Thus, considering the possibility of alternating yield at any section at which the maximum and minimum moments are $\mathcal{M}_j^{\max}$ and $\mathcal{M}_j^{\min}$ respectively, the load factor λ given by

$$\lambda (\mathcal{M}_j^{\max} - \mathcal{M}_j^{\min}) = 2(M_y)_j \tag{5.18}$$

will be an upper bound on the shakedown limit λ_s for the structure. The minimum value λ_A of all values of λ so calculated will also be an upper bound on the shakedown limit.

Considering incremental collapse, let any static collapse mechanism have rotations θ_j at sections h_j, the plastic moments at these sections being M_{pj}. The plastic moments are positive scalar quantities, but the rotations θ_j are assumed to have signs corresponding to the sign convention for the maximum and minimum elastic moments $\lambda \mathcal{M}_j^{\max}$ and $\lambda \mathcal{M}_j^{\min}$. Suppose now that there exists a set of residual moments with values m_j at sections h_j such that

$$\left. \begin{array}{l} m_j + \lambda \mathcal{M}_j^{\max} = M_{pj} \quad \text{if } \theta_j > 0 \\ m_j + \lambda \mathcal{M}_j^{\min} = -M_{pj} \text{ if } \theta_j < 0 \end{array} \right\} \tag{5.19}$$

Then incremental collapse would take place according to the postulated mechanism at the load factor λ, assuming that it did not occur by some other mechanism at a lower load factor. Hence, λ is an upper bound on the shakedown limit.

The unknown moments m_j in Eqn (5.19) may be eliminated by considering the virtual work equation which arises when these residual moments (in equilibrium with zero external loads) are associated with the hinge rotations θ_j. Since

$$\sum_j m_j \theta_j = 0 \tag{5.20}$$

it follows from (5.19) that

$$\lambda \sum_j \left\{ \begin{array}{l} \mathcal{M}_j^{\max} \\ \mathcal{M}_j^{\min} \end{array} \right\} \theta_j = \sum_j M_{pj} |\theta_j| \tag{5.21}$$

where $\mathcal{M}_j^{\max}$ is taken when θ_j is positive, and $\mathcal{M}_j^{\min}$ when θ_j is negative.

Any collapse mechanism may thus be used to obtain an upper bound on the shakedown limit by substituting the appropriate hinge rotations in Eqn (5.21). If λ_I is the minimum value obtained by considering all possible mechanisms, the actual shakedown limit λ_s is the smaller of the two load factors λ_I and λ_A, obtained by considering incremental collapse and alternating yield respectively.

5.4 Methods of shakedown analysis

The derivation of the upper bound λ_A obtained by considering alternating yield is straightforward and requires no further elaboration. The incremental collapse limit λ_I may be obtained by suitable adaptations of any of the methods described in Chapters 1 and 2 for static collapse. Simple problems may be solved by inspection, either analytically or graphically. In some problems, it is possible systematically to explore all possible mechanisms, thus ensuring that the correct mechanism is chosen. A combined mechanism method may be used as a means of exploring all likely mechanisms when more extensive structures are involved. In place of the final static check, carried out for static loading to ensure that the correct mechanism has been chosen, it is necessary, for variable repeated loading, to derive a residual set of moments which is shown to satisfy the inequalities (5.2). These methods of analysis are all suited to hand calculations and will now be illustrated.

A method of systematic elimination of unknowns from a set of linear inequalities, similar to that used for estimating static collapse loads, is more suitable if a digital computer is to be used, but will not be dealt with here (see Heyman[7]).

Solution by inspection

Consider the two-span example in Fig. 5.4(a) in which the loading alternates between $\lambda_1 = \lambda_2 = \lambda$ and $\lambda_1 = 0, \lambda_2 = \lambda$. The elastic moments under these two loading conditions are given in the first two rows of Table 5.3, hogging moments being accounted positive. Incremental collapse will, if it occurs, be

Table 5.3   Calculation of shakedown limit for beam in Fig. 5.4(a)

Section	B	C	D
Bending moments, $\lambda_1 = \lambda_2 = \lambda$	$-\frac{5}{32}\lambda M_p$	$\frac{3}{16}\lambda M_p$	$-\frac{5}{32}\lambda M_p$
Bending moments, $\lambda_1 = 0, \lambda_2 = \lambda$	$\frac{3}{64}\lambda M_p$	$\frac{3}{32}\lambda M_p$	$-\frac{13}{64}\lambda M_p$
Rotations, mechanism Fig. 5.4(b)		θ	-2θ

by the mechanism in Fig. 5.4(b), giving the hinge rotations shown in the third row of Table 5.3. Applying Eqn (5.21),

$$(\tfrac{3}{16}\lambda M_p)\theta + (-\tfrac{13}{64}\lambda M_p)(-2\theta) = 3M_p\theta$$

whence $\lambda = \lambda_I = \tfrac{96}{19} = 5\!\cdot\!06$

agreeing with the step-by-step solution given earlier. The maximum range of moment occurs at B, so that the minimum load factor for alternating yield is given by

$$(\tfrac{3}{64} + \tfrac{5}{32})\lambda M_p = 2M_y$$

whence $\lambda = \lambda_A = \dfrac{128}{13}\dfrac{M_y}{M_p} = 8\!\cdot\!57,$

assuming $M_p/M_y = 1\!\cdot\!15$. Hence the shakedown limit is controlled by incremental collapse, that is, $\lambda_s = 5\!\cdot\!06$.

Graphical solution

Suppose a concentrated load W can act anywhere within the span of a fixed-ended beam AB [Fig. 5.5(a)]. The maximum and minimum elastic moments (hogging moments positive) are readily found to be as illustrated in Fig. 5.5(b), where

$$\mathscr{M}^{\max} = \frac{4}{27}\frac{(L - 3x)^3}{(L - 2x)^2}\,W, \quad 0 \leqslant x \leqslant \frac{L}{3}$$

$$\mathscr{M}^{\max} = 0, \qquad\qquad \frac{L}{3} \leqslant x \leqslant \frac{L}{2}$$

$$\mathscr{M}^{\min} = \frac{2x^2(L - x)^2}{L^3}\,W, \quad 0 \leqslant x \leqslant \frac{L}{2}$$

Inspection shows that the residual moment line a_2b_2 lies in such a position that incremental collapse can take place by the mechanism in Fig. 5.5(c), whence in Fig. 5.5(b),

$$a_2a_1 = c_1c_2 = M_p$$

Hence $\tfrac{4}{27}W_sL + \tfrac{1}{8}W_sL = 2M_p$

and $\quad W_s = \dfrac{432}{59}\dfrac{M_p}{L} = 7\!\cdot\!32\dfrac{M_p}{L}$

The residual moment is of value

$$Aa_2 = \tfrac{4}{27}\,W_sL - M_p = \tfrac{5}{59}\,M_p = 0\!\cdot\!085M_p$$

At loads slightly in excess of the shakedown limit W_s, the central plastic hinge at C deforms when the load is at midspan. The end hinges at A and B deform when the load is $L/3$ from A and B respectively, these being the positions for maximum hogging moments at the ends.

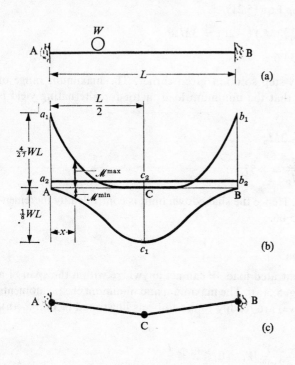

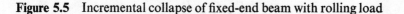

Figure 5.5   Incremental collapse of fixed-end beam with rolling load

It is readily confirmed that alternating yield does not occur. Static collapse occurs at $W_p = 8(M_p/L)$ with the load at midspan.

Exploration of possible mechanisms

We consider the uniform rectangular fixed base portal frame in Fig. 5.6(a), subjected to a central vertical load varying between zero and $\lambda_V (M_p/L)$ and a horizontal load at beam level varying between zero and $\lambda_H(M_p/L)$. It is assumed that the load variations can take place independently, and we shall explore the problem for all positive combinations of λ_V and λ_H. The sign convention for moments and hinge rotations is that tension is positive on the tops of beams and on the left-hand sides of columns. In quoting moments and rotations at joints B and D [Fig. 5.6(a)], the sign convention for the beam is used.

The maximum and minimum elastic moments are given in the first two rows of Table 5.4, and the hinge rotations for the three possible incremental mechanisms 1, 2, and 3 [Fig. 5.6(b)] are shown in the succeeding rows. The application of Eqn (5.21) to mechanism 1 gives

$$(\tfrac{14}{84}\lambda_V M_p)\theta + (\tfrac{28}{84}\lambda_V M_p)2\theta + (\tfrac{14}{84}\lambda_V M_p + \tfrac{36}{84}\lambda_H M_p)\theta = 4M_p\theta$$

that is, for *mechanism 1* $\tfrac{1}{4}\lambda_V + \tfrac{3}{28}\lambda_H = 1$ (5.22)

Similarly, for mechanisms 2 and 3 respectively,

mechanism 2 $\frac{13}{72}\lambda_V + \frac{1}{3}\lambda_H = 1$ (5.23)

mechanism 3 $\frac{1}{16}\lambda_V + \frac{1}{2}\lambda_H = 1$ (5.24)

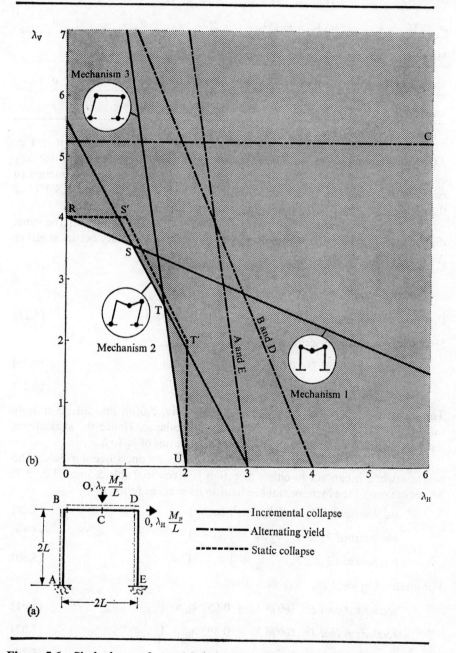

Figure 5.6   Shakedown of portal frame with unidirectional horizontal load

Table 5.4   Calculation of shakedown limit for portal frame in Fig. 5.6

Section	A	B	C	D	E
$\mathcal{M}^{\max}/M_p$	$\frac{48}{84}\lambda_H$	$\frac{14}{84}\lambda_V$	0	$\frac{14}{84}\lambda_V + \frac{36}{84}\lambda_H$	$\frac{7}{84}\lambda_V + \frac{48}{84}\lambda_H$
$\mathcal{M}^{\min}/M_p$	$-\frac{7}{84}\lambda_V$	$-\frac{36}{84}\lambda_H$	$-\frac{28}{84}\lambda_V$	0	0
Mechanism 1		θ	-2θ	θ	
Mechanism 2	θ		-2θ	2θ	θ
Mechanism 3	θ	$-\theta$		θ	θ

These three equations may be expressed graphically as shown in Fig. 5.6(b), and define a permissible region bounded by RSTU within which any combination λ_V and λ_H will not cause incremental collapse. It is of interest to compare this boundary for incremental collapse with the boundary RS'T'U of the permissible region for collapse under a single application of loads.

The conditions for alternating plasticity may be shown on the same diagram. Assuming a shape factor of 1·15, alternating plasticity occurs at either A or E (Table 5.4) when

$$\frac{7}{84}\lambda_V M_p + \frac{48}{84}\lambda_H M_p = 2\left(\frac{M_p}{1\cdot 15}\right)$$

that is, for *sections A and E* $0\cdot048\,\lambda_V + 0\cdot328\,\lambda_H = 1$ (5.25)

Similarly, for the other critical sections,

sections B and D $0\cdot096\,\lambda_V + 0\cdot246\,\lambda_H = 1$ (5.26)

section C $0\cdot192\,\lambda_V = 1$ (5.27)

These equations are shown as dotted lines in Fig. 5.6(b), and all lie entirely outside the permissible region for incremental collapse. Hence the shakedown limit is controlled by incremental collapse for all ratios of λ_V to λ_H.

In this example, the horizontal load acts in one direction only. The solution when it can act in either direction is given in Table 5.5 and Fig. 5.7. The equations of the incremental mechanism lines are as follows.

mechanism 1 $\frac{1}{4}\lambda_V + \frac{3}{14}\lambda_H = 1$ (5.28)

mechanisms 2 and 4 $\frac{13}{72}\lambda_V + \frac{1}{3}\lambda_H = 1$ (5.29)

mechanisms 3 and 5 $\frac{1}{16}\lambda_V + \frac{1}{2}\lambda_H = 1$ (5.30)

For alternating yield they are as follows.

sections A and E $0\cdot048\,\lambda_V + 0\cdot657\,\lambda_H = 1$ (5.31)

sections B and D $0\cdot096\,\lambda_V + 0\cdot493\,\lambda_H = 1$ (5.32)

section C $0\cdot192\,\lambda_V = 1$ (5.33)

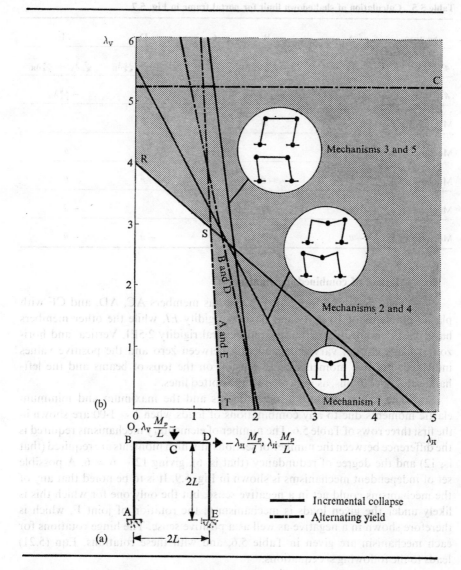

Figure 5.7 Shakedown of portal frame with alternating horizontal load

The boundary RST to the permissible region [Fig. 5.7(b)] is controlled by incremental collapse by mechanism 1 from R to S ($\lambda_H/\lambda_V < 0.46$), and by alter­nating yield at the bases A and E from S to T ($\lambda_H/\lambda_V > 0.46$). It is not unexpected that alternating yield should control the shakedown limit for higher values of the alternating horizontal load, but the action of smaller alternating horizontal loads in inducing a pure-beam-mechanism type of incremental collapse is of interest, and of possible significance for the effect of wind loading on design.

Table 5.5 Calculation of shakedown limit for portal frame in Fig. 5.7

Section	A	B	C	D	E
$\mathscr{M}^{max}/M_p$	$\frac{48}{84}\lambda_H$	$\frac{14}{84}\lambda_V + \frac{36}{84}\lambda_H$	0	$\frac{14}{84}\lambda_V + \frac{36}{84}\lambda_H$	$\frac{7}{84}\lambda_V + \frac{48}{84}\lambda_H$
$\mathscr{M}^{min}/M_p$	$-\frac{7}{84}\lambda_V - \frac{48}{84}\lambda_H$	$-\frac{36}{84}\lambda_H$	$-\frac{28}{84}\lambda_V$	$-\frac{36}{84}\lambda_H$	$-\frac{48}{84}\lambda_H$
Mechanism 1		θ	-2θ	θ	
Mechanism 2	θ		-2θ	2θ	θ
Mechanism 3	θ	$-\theta$		θ	θ
Mechanism 4	$-\theta$	2θ	-2θ		$-\theta$
Mechanism 5	$-\theta$	θ		$-\theta$	$-\theta$

Method of combined mechanisms

The two-storey frame in Fig. 5.8 has members AC, AD, and CF with plastic moments of 10 units and flexural rigidity *EI*, while the other members have plastic moments of 20 units and flexural rigidity 2·5EI. Vertical and horizontal loads can all vary independently between zero and the positive values indicated. Positive moments cause tension on the tops of beams and the left-hand sides of columns, as indicated by the dotted lines.

The values of the plastic moments and the maximum and minimum elastic moments due to any combinations of loads when $\lambda = 1·00$ are shown in the first three rows of Table 5.6. The number of elementary mechanisms required is the difference between the number of sections at which moments are required (that is, 12) and the degree of redundancy (that is, 6), giving $12 - 6 = 6$. A possible set of independent mechanisms is shown in Fig. 5.9. It is to be noted that any of the mechanisms could act in a negative sense, but the only one for which this is likely under the given loads is mechanism 6, the rotation of joint F, which is therefore shown in a negative as well as a positive sense. The hinge rotations for each mechanism are given in Table 5.6, and with these rotations, Eqn (5.21) leads to the following six equations.

mechanism 1	$13·10\lambda\theta = 40\theta,$	$\lambda = 3·06$
mechanism 2	$30·73\lambda\theta = 80\theta,$	$\lambda = 2·60$
mechanism 3	$12·95\lambda\theta = 40\theta,$	$\lambda = 3·09$
mechanism 4	$44·44\lambda\theta = 80\theta,$	$\lambda = 1·80$
mechanism 5	$14·86\lambda\theta = 50\theta,$	$\lambda = 3·37$
mechanisms 6a and b	$14·86\lambda\theta = 50\theta,$	$\lambda = 3·37$

It is to be noted that, in contrast to the situation under static loading, the joint rotations yield a finite load factor λ. This is because the incremental rotation of a joint is physically possible under variable repeated loading. (Under

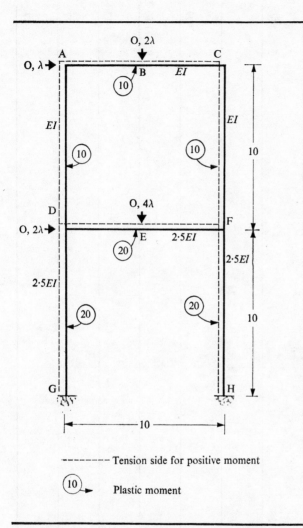

Figure 5.8 Variable loading of two-storey frame

static loading, a joint to which an external moment was applied would, of course, give a joint rotation mechanism with a finite load factor.)

The smallest load factor among the elementary mechanisms ($\lambda = 1\cdot80$) is for the sway of the lower storey (mechanism 4). The effect may be tried of combining this with mechanism 2 ($\lambda = 2\cdot60$) together with the rotation of joint D (mechanism 5), giving the combined mechanism in Fig. 5.10(a). The combined virtual work equation is then

$$(30\cdot73 + 44\cdot44 + 14\cdot86)\,\lambda\theta - (3\cdot76 + 10\cdot51)\,\lambda\theta - (2\cdot75 + 8\cdot47)\,\lambda\theta$$
$$= (80 + 80 + 50)\theta - (20)2\theta - (20)2\theta \quad (5.34)$$

that is, $64\cdot54\lambda\theta = 130\theta$

$$\lambda = 2\cdot015$$

10—P.T.S.

Table 5.6 Calculation of shakedown limit for frame in Fig. 5.8

	M_{AC}	M_B	M_{CA}	M_{DA}	M_{DF}	M_{DG}	M_E	M_{FC}	M_{FD}	M_{FH}	M_{GD}	M_{HF}	λ
Plastic moment M_P	10	10	10	10	20	20	20	10	20	20	20	20	
Maximum elastic moment $\mathscr{M}^{max}$	1·00	0·18	3·74	2·61	3·76	2·75	0·11	3·86	14·27	0·23	11·86	13·12	1·00
Minimum elastic moment $\mathscr{M}^{min}$	−2·74	−4·18	0	−1·60	−10·51	−8·47	−6·35	−0·35	0	−10·99	−1·37	−0·11	1·00
Mechanism 1 Upper beam collapse	θ	-2θ	θ										3·06
Mechanism 2 Lower beam collapse					θ		-2θ		θ				2·60
Mechanism 3 Upper storey sway	$-\theta$		θ	θ				θ					3·09
Mechanism 4 Lower storey sway						$-\theta$				$-\theta$		θ	1·80
Mechanism 5 Clockwise rotation, joint D				$-\theta$	$-\theta$	θ							3·37
Mechanism 6a Clockwise rotation, joint F								$-\theta$	$-\theta$	θ			3·37
Mechanism 6b Anticlockwise rotation, joint F								θ	$-\theta$	$-\theta$			3·37

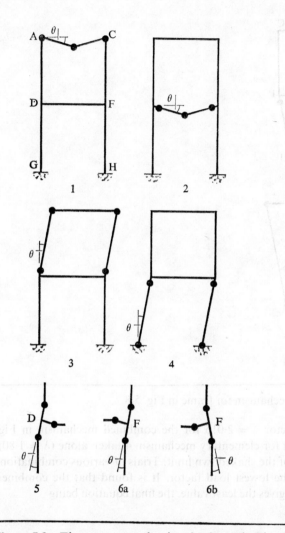

Figure 5.9 Elementary mechanism for frame in Fig. 5.8

The second bracket on the left-hand side of Eqn (5.34) represents the cancellation of terms corresponding to the left-hand side of Eqn (5.21) due to the elimination of the hinge rotations at end D of member DF. These cancelled rotations appear as positive and negative values in the appropriate rows (for mechanisms 2 and 5) and column (M_{DF}) of Table 5.6. Similarly, the third bracket in Eqn (5.34) represents the cancellation terms for the elimination of rotations at end D of member DG. Corresponding terms for cancelled plastic hinge work occur on the right-hand side of Eqn (5.34). It may be noted that all the cancelling terms may be derived directly from Table 5.6 by noting where hinge rotations of opposite sign are combined when adding the mechanisms.

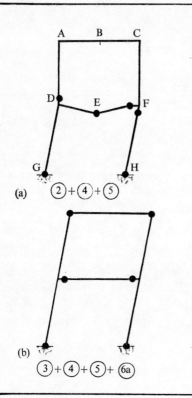

(a)   (2)+(4)+(5)

(b)

(3)+(4)+(5)+(6a)

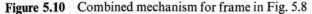

Figure 5.10   Combined mechanism for frame in Fig. 5.8

Since the load factor $\lambda = 2 \cdot 015$ for the combined mechanism in Fig. 5.10(a) is greater than that for elementary mechanism 4 taken alone ($\lambda = 1 \cdot 80$), it is not the correct value of the shakedown limit. Trials of various combinations may be made to obtain the lowest load factor. It is found that the combined mechanism in Fig. 5.10(b) gives the least value, the final equation being

$$56 \cdot 24 \lambda \theta = 100 \theta$$

that is, $\lambda = 1 \cdot 78$

Surprisingly, the elementary beam mechanism 2 is not involved, and this emphasizes the importance of carrying out a check to ensure that all changes of moment can be accommodated elastically. The procedure is to derive a set of residual moments satisfying inequalities (5.2).

Such a set is shown for the present frame in Fig. 5.11. The residual moment at G is derived by noting that a plastic hinge ($M_{GD} = 20$) occurs when the loading is such as to produce maximum positive moment. Since the load factor is $1 \cdot 78$, the residual moment is therefore $20 - 1 \cdot 78(11 \cdot 86) = -1 \cdot 10$. Similarly, the residual moment at H is found to be $-3 \cdot 35$. The collapse mechanism in Fig. 5.10(b) leaves the bending moments M_{DG} and M_{FH} indeterminate—

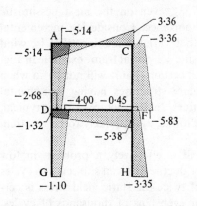

Figure 5.11 Residual moments for frame in Fig. 5.8 undergoing incremental collapse as in Fig. 5.9

this is because it is an incomplete mechanism, there being six plastic hinges and six degrees of redundancy. If this mechanism were incorrect, there would be a likelihood of a negative plastic hinge at F in FH, and so the residual moment at this section is calculated on this assumption. The residual moment m_{FH} thus becomes $-20 + 1{\cdot}78 \times 10{\cdot}99 = -0{\cdot}45$. Since the residual moments are in equilibrium with zero external loads, the elementary mechanism for sway in the lower storey (mechanism 4, Fig. 5.9) may be used to obtain a relationship between the four residual moments as follows:

$$-m_{\mathrm{DG}} - m_{\mathrm{FH}} + m_{\mathrm{GD}} + m_{\mathrm{HF}} = 0$$

Hence $m_{\mathrm{DG}} = -4{\cdot}00$. The remaining residual moments are now fully determinate, and are derived by continuing the procedure. It is readily confirmed that the maximum and minimum elastic moments from Table 5.6 at load factor 1·78, when added to the residual moments in Fig. 5.11, produce moments nowhere greater than the plastic moments. Hence the shakedown conditions are satisfied, and $\lambda_s = 1{\cdot}78$.

5.5   Significance of shakedown in design

While incremental collapse is theoretically possible in most structures at a load factor λ_s less than the load factor for static collapse λ_p, the importance of the effect of variable repeated loading in design depends on the probability of a sufficient number of load variations occurring above the shakedown limit for significant permanent deformations to be induced. A study of the frequencies of varying intensities of load in relation to the design intensity for both floor and wind loading was made with reference to this problem by Horne.[8] Although the data on which this study is based is limited, it seems clear enough that, considering incremental collapse, repeated floor loading is unlikely to be of importance in a structure designed on the basis of static collapse at load factor λ_p when, for that structure, $\lambda_s > 0.66\lambda_p$. When $\lambda_s > 0.75\lambda_p$, even on the most pessimistic assumptions, repeated loading is of no importance. Considering incremental collapse due to wind loads, it is unlikely that variable loading (that is, wind first from one direction, then from the opposite, see portal frame example in Fig. 5.7) will be important when $\lambda_s > 0.60\lambda_p$. and certain that it will not be so when $\lambda_s > 0.64\lambda_p$. While it is possible to postulate structures having incremental collapse shakedown limits less than these values, such structures are not typical, and for realistic structures and loading conditions, λ_s is unlikely to be less than $0.75\lambda_p$.

The importance of alternating yield is effectively a problem in low cycle fatigue. The fatigue life of structures of ductile material subjected to cycles which involve a stress range of the order of twice the static yield stress is certainly measured in hundreds, thousands, or even tens of thousands of cycles, except to the extent that joints can lead to severe stress-raiser problems. Apart from its importance for joint detailing, alternating yield is only likely to be more critical than incremental collapse when the design is dominated by wind loading, and by using wind loading frequency curves, Horne shows that alternating yield is most unlikely to cause failure when $\lambda_s > 0.56\lambda_p$.

The general conclusion appears to be that, as far as design for safety is concerned, variable repeated loads are unlikely to be of importance unless conditions similar to those of ordinary fatigue are approached. A study of shakedown loads for crane girders[9] indicates that, even when variable loading is a basic feature of the design situation, design may be carried out on the basis of simple plastic collapse unless there is a high probability of the girder carrying its full design load for a large proportion of the time. This is a fortunate conclusion, since the calculation of shakedown loads, in contrast to that of failure loads, involves a knowledge of the relative elastic stiffnesses of the members. Direct design methods, and in particular minimum weight procedures, are not generally practicable, although Heyman[10] has shown how very simple cases may be dealt with. A knowledge of shakedown theory is, however, required as a check on the behaviour of a structure under repeated loading if, after designing on the basis of static loading, there are reasons for believing that the variable repeated loading conditions are particularly severe.

Bibliography

1   Royles, R. Low endurance fatigue behaviour of mild steel beams in reversed bending. *J. Stress Analysis*, 1966, **1**, 239.

2   Bleich, H. Über die Bemessung statisch unbestimmter Stahltragwerke unter Berücksichtigung des elastisch-plastischen Verhaltens des Baustoffes. *Bauingenieur*, 1932, **19/20**, 261.

3   Melan, E. Theorie statisch unbestümmter Systeme. Preliminary publi-
     cation of the 2nd Congress of the International Association of Bridge
     and Structural Engineering, Berlin, 1936, 43.
4   Neal, B. G. The behaviour of framed structures under repeated loading.
     Quart. J. Mech. Appl. Math., 1951, **4**, 78.
5   Neal, B. G. Plastic collapse and shakedown theorems for structures
     of strain-hardening material. *J. Aero. Sci.*, 1950, **17**, 297.
6   Neal, B. G. *The Plastic Methods of Structural Analysis*. Chapman and
     Hall, London, 1956.
7   Heyman, J. Automatic analysis of steel framed structures under fixed
     and varying loads. *Proc. Instn civ. Engrs*, 1959, **12**, 39.
8   Horne, M. R. *The effect of variable repeated loads in buildings designed
     by plastic theory*. International Association of Bridge and Structural
     Engineering, Zürich, 1954, 53.
9   Horne, M. R. and J. M. Davies. Repeated loading in the plastic design
     of crane girders. *Engineer*, 1963, **216**, 1053.
10   Heyman, J. Plastic design of beams and plane frames for minimum
     material consumption. *Quart. App. Math.*, 1951, **8**, 373.

Problems

5.1   A uniform continuous beam ABC in which AB = BC = L is simply
      supported at A, B, and C. Show that the elastic hogging moment over
      the support B due to a concentrated load W distance a from A (where
      $0 < a < L$) is $Wa(L^2 - a^2)/4L^2$. Hence obtain an expression for the
      shakedown limit W in terms of the plastic moment M_p of the beam
      when loads W_D and W_E act independently at D and E, where AD = EC
      = a, the variations of W_D and W_E each being between the limits 0 and
      W.

5.2   A uniform continuous beam ABCD where AB = CD = L, BC = $2L$
      is simply supported at A, B, C, and D, and a moving load of magnitude
      W travels between B and C. If, when the load is at a distance a from B
      the elastic moment induced at B in the absence of residual moments is
      $Wa(2L - a)(5L - 2a)/16L^2$, find the shakedown limit of W.

5.3   A uniform beam ABCDE where AB = BC = CD = DE = L is simply
      supported at A, B, C, D, and E. Four equal loads, applied at the centre
      of each span, can vary independently between 0 and W. Determine the
      maximum value of W as controlled (a) by alternating yield, assuming
      an elastic range of $1 \cdot 6M_p$ where M_p is the plastic moment, and (b) by
      incremental collapse. The elastic moments (hogging moments positive)
      induced at B, C, and D by a load W at the centre of span AB are $0 \cdot 101WL$,
      $-0 \cdot 027WL$, and $0 \cdot 006WL$ respectively, and by a load W at the centre
      of span BC they are $0 \cdot 074WL$, $0 \cdot 080WL$, and $-0 \cdot 019WL$ respectively.

5.4   The rigid-jointed, fixed-base frame in Fig. E5.1 has a uniform section throughout with plastic moment 200 units. The loads W_A, W_B, and W_C vary independently between the following limits

$$W_A \quad 0, 100$$
$$W_B \quad 0, 100$$
$$W_C \quad 0, 50$$

The elastic moments induced at sections 1 to 10 (Fig. E5.1) by loads $W_A = 100$, $W_B = 100$, and $W_C = 50$ are as follows, where clockwise moments acting on the ends of members are positive and moments at sections 3 and 8 are positive when sagging.

	1	2	3	4	5	6	7	8	9	10
$W_A = 100$	24	−56	127	90	−54	−32	−36	−15	−6	0
$W_B = 100$	0	6	−15	36	54	32	−90	127	56	−24
$W_C = 50$	−71	50	3	44	−76	−82	44	−3	50	−71

Determine the greatest load factor which may be applied to the given loading without the occurrence of incremental collapse, and the associated residual moments in the frame.

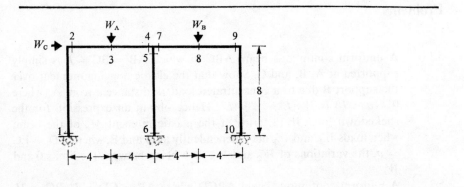

Figure E5.1

5.5   The loads W_A, W_B, and W_C acting on the frame described in Example 5.4 (Fig. E5.1) vary independently between limits as follows

$$W_A \quad 0, 100$$
$$W_B \quad 0, 100$$
$$W_C \quad -50, 50$$

Determine the shakedown load factor as controlled by the incidence of (a) incremental collapse, (b) alternating yield. The plastic moment is of uniform value 200 units, and the moment causing yield in the extreme fibres is 175 units.

Chapter 6
Stability

6.1   Introduction

Plastic collapse loads as discussed in previous chapters are idealizations to the failure loads of elastic–plastic structures. In these idealizations, the effects of deformations on the equations of equilibrium are ignored, the assumed geometry being in each case that of the undeformed structure. Strictly speaking, therefore, the collapse load refers to infinitely small departures from the undeformed states with infinitely small plastic hinge rotations. Actual structures undergo elastic deformations before any plastic deformation occurs, followed by still larger deformations as the degree of plasticity increases. Whether or not these deformations are sufficient to affect significantly the failure load as compared with the rigid–plastic collapse load depends on the particular structure, and also on the loading pattern. Very slender structures may approach a condition of failure due to elastic instability well before the theoretical collapse load given by rigid–plastic theory is reached, but for a wide range of structures the failure loads are found to be very close to the rigid–plastic collapse values.

It is obviously necessary to establish criteria that will determine when the effect of change of geometry must be taken into account. A full treatment of this topic is well outside the scope of this volume, but the influence of various structural parameters and appropriate methods for the estimation of failure loads are discussed.

Change of geometry may be classified according to whether their influence is primarily:

(a)   in the structure as a whole ('overall stability')
(b)   within the length of a member relative to the points of support or attachment of the member ('member stability');
(c)   within the member, affecting its cross-sectional shape ('local stability').

An example of overall stability is the effect of sway deformation on the failure load of a multi-storey building frame. Member stability is the factor being taken into account when formulae are used limiting the stresses in columns or slender beams, usually depending on the 'slenderness ratio' (ratio of length to least radius of gyration). Local stability is violated when a flange or web plate buckles, and, if this is to be avoided, limitations must be placed on the plate width to thickness ratios.

Although methods of dealing with these problems appear to vary quite widely, the underlying physical behaviour is common. This behaviour involves the occurrence of finite deformations which are sufficient to affect the equilibrium conditions, so that changes occur in the internal forces and stresses induced by the external loads, the overall tendency being for the stresses to be increased. This causes further deformation, and ultimately collapse can occur at loads less than those derived by ignoring the deformations.

The general features of stability in relation to plastic collapse are discussed in Section 6.2, using as an example the behaviour of an eccentrically loaded column. Methods of obtaining upper and lower bounds on failure loads and various methods for estimating actual values are then illustrated by reference to a portal frame problem in Sections 6.3 and 6.4.

6.2   Eccentrically loaded column

The initially straight steel column AB in Fig. 6.1(a) is of rectangular cross section of side $d = 0.1$ m and of length $L = 2$ m. The axis of the column is supported initially in a vertical position by the rigid base at A, and at the top B a vertical load P acts at an eccentricity of $e = d = 0.1$ m. The steel is assumed to have an elastic–pure plastic stress–strain relationship with an elastic modulus $E = 200 \times 10^9$ N/m^2 and a yield stress $\sigma_y = 250 \times 10^6$ N/m^2. The second moment of area I is $(0.1)^4/12 = 8.33 \times 10^{-6}$ m^4, and hence the flexural rigidity EI is 1.667×10^6 Nm^2.

The behaviour as long as the column behaves elastically may be derived in the usual way. The deflection y_B of the top section is given by the well-known secant formula

$$y_B = e \left\{ \sec \frac{\pi}{2} \sqrt{\left(\frac{P}{P_E}\right)} - 1 \right\} \tag{6.1}$$

where P_E is the Euler buckling load of the column. Since the column is free to sway at the top,

$$P_E = \frac{\pi^2 EI}{4L^2} = 1.027 \times 10^6 \text{ N}$$

whence $y_B = 0.1\{\sec (1.55 \times 10^{-3} \sqrt{P}) - 1\}$ \hfill (6.2)

This elastic load-deflection curve is shown graphically by OGH in Fig. 6.2. It rises asymptotically to the Euler critical load P_E, although small deflection theory on which Eqn (6.1) is based ceases to be a reasonable approximation after the deflection rises above about $0.1L$ (that is, 0.2 m).

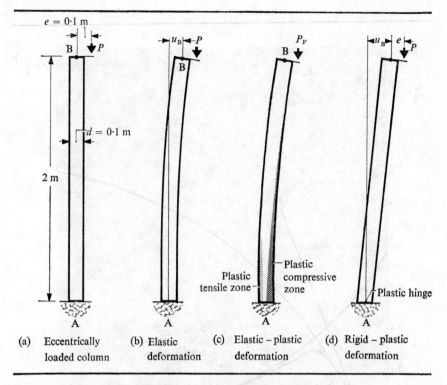

Figure 6.1 Elastic–plastic behaviour of eccentrically loaded column

The elastic solution ceases to apply after the yield stress has been reached on the concave face of the column at A. This occurs when

$$\frac{P}{A} + \frac{P(e + y_B)}{Z} = \sigma_y \tag{6.3}$$

where A is the area of cross section ($d^2 = 0.01$ m^2), and Z is the elastic modulus ($d^3/6 = 1.667 \times 10^{-4}$ m^3). Substituting for y_B from Eqn (6.1) and solving, the load P_Y at first yield is found to have the value

$$P_Y = 0.212 \times 10^6 \text{ N}$$

represented by point Y on the load-deflection curve (Fig. 6.2).

It is possible, by following the formation first of compressive, then of tensile plastic zones [Fig. 6.1(c)], to derive flexural equations for the various parts of the column. When this is done[1] the load-deflection curve is found to rise to a peak and then to descend as shown by YFM in Fig. 6.2. This is a laborious process, but establishes the theoretical failure load

$$P_F = 0.325 \times 10^6 \text{ N}$$

It is desirable to find some simpler means of obtaining an estimate of the failure load. As has already been seen, by ignoring *yield* (that is, by assuming

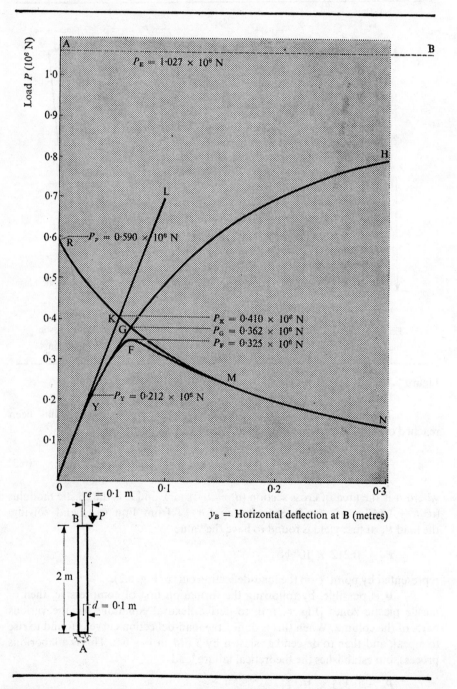

Figure 6.2 Load-deflection curves for eccentrically loaded column

indefinite elastic behaviour), we obtain the *elastic deformation curve* 0GH. A complementary step which may be taken is to calculate the behaviour by ignoring *elasticity* (that is, by assuming rigid–plastic behaviour). This involves a mechanism with a plastic hinge at A,† as shown in Fig. 6.1(d). Let $P = nP_s$, where $P_s(= d^2\sigma_y)$ is the *squash load* for the column. Then (Chapter 3) the plastic moment is of value $(1 - n^2)(d^3\sigma_y/4)$, and equating this to the externally applied moment,

$$(e + y_B)nd^2\sigma_y = (1 - n^2)d^3\sigma_y/4$$

Hence $e + y_B = \left(\dfrac{1 - n^2}{n}\right)\dfrac{d}{4}$ \hfill (6.4)

Substituting $P = nP_s = n(0{\cdot}1)^2 \times 250 \times 10^6 = n \times 2{\cdot}5 \times 10^6 \text{ N}$ for   the column under consideration, Eqn (6.4) gives the *rigid–plastic mechanism curve* RMN in Fig. 6.2. It is to be noted that the elastic–plastic curve OYFM leaves the elastic deformation curve at Y and joins the mechanism curve at M, so that the failure load P_F must lie below, but necessarily closely related to, the load P_G at the point of intersection of these curves. In the present case, it is found that

$$P_G = 0{\cdot}362 \times 10^6 \text{ N}$$

so that $\dfrac{P_F}{P_G} = 0{\cdot}898$

While the intersection load G unfortunately always gives a high (upper bound) estimate of the failure load, it is readily calculated, and has for this reason been investigated extensively by Murray[2,3] as applied to compression members contained in triangulated frames. Murray obtained his failure loads P_F experimentally, and calculated his load P_G by allowing for elastic deflections arising from imperfections. He found that P_F/P_G varied from 0·77 to 0·98. A similar method may be used for compresssion members carrying lateral loads ('beam columns'), in which case the moments due to the lateral loads are taken into account when calculating both the elastic and the rigid–plastic curves.

An intersection point which is somewhat easier to calculate than point G in Fig. 6.2 is point K, where 0KL is the *linear elastic* load–deflection relation. This is obtained by ignoring the effect of change of geometry on the elastic behaviour of the column, which is therefore assumed to sustain a uniform moment Pe throughout its length. This gives

$$y_B = e\left(\frac{PL^2}{2EI}\right)$$

giving $y_B = 1{\cdot}2 \times 10^{-6}P$

---

† Since the column, when rigid, has a uniform moment Pe, the moment is uniform and the hinge can occur anywhere. The same result is obtained whatever position is assumed for the hinge.

Hence it is found that

$$P_K = 0.410 \times 10^6 \text{ N}$$

and $\dfrac{P_F}{P_K} = 0.794$

The loads P_G and P_K are both *upper bounds* on the failure load P_F. More remote upper bounds are the elastic critical load $P_E = 1.027 \times 10^6 \text{ N}$ (point A in Fig. 6.2), and the rigid–plastic failure load P_p for zero deformation, represented by point R. This latter load is obtained by putting $y_B = 0$ in Eqn (6.4), whence for this column $n = \sqrt{5} - 2 = 0.236$, and

$$P_p = 0.590 \times 10^6 \text{ N}$$

Hence $\dfrac{P_F}{P_p} = 0.550$

The only *lower bound* on the failure load that can readily be calculated is the load P_Y at first yield, that is, $0.212 \times 10^6 \text{ N}$.

6.3   Portal frame

We consider the fixed-base portal frame in Fig. 6.3(a), loaded as shown with a uniform plastic moment $M_p = 2WL$. In discussing the effect of defor- mations on collapse loads, it is necessary to know the flexural rigidity for bending in the plane of the frame, and this will be taken as $EI = 5WL^2$. Hence $EI = 2.5M_pL$. Assuming a shape factor of 1.15 and a ratio of radius of gyration r to depth d of 0.4, it follows that

$$1.00 = \frac{2.5M_pL}{EI} = \frac{2.5\left(1.15\dfrac{2I}{d}\sigma_y\right)L}{EI} = 2.3\left(\frac{\sigma_y}{E}\right)\left(\frac{L}{r}\right)$$

A typical value of E/σ_y for mild steel is 900, so that L/r is of the order of 400. The frame is thus an unrealistically slender one, but this case has been chosen on purpose to illustrate clearly the effect of change of geometry. A single-storey frame with a stiffness more in the practical range would show much less reduc- tion of carrying capacity due to deformations, although appreciable reductions are possible in practical multi-storey frames. Since the frame is so slender the effect of axial load on the plastic moment may be neglected.

The frame in Fig. 6.3(a) has the same rigid–plastic collapse load as that considered in detail in Chapter 1 (Fig. 1.9) since the central beam and horizontal sway loads are identical. The collapse mode is as shown in Fig. 6.3(b), with load factor for infinitely small deformations θ of $\lambda_p = 2.40$. This load factor is evidently an upper bound on the elastic–plastic collapse load factor λ_F.

In order to determine the rigid–plastic mechanism line corresponding

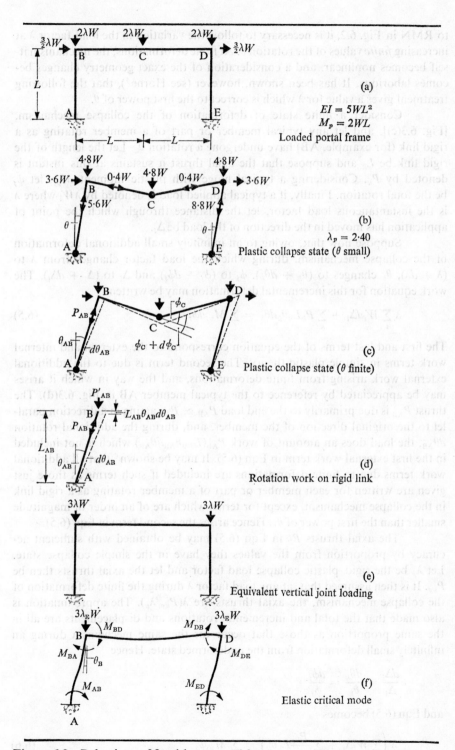

(a)

$EI = 5WL^2$
$M_p = 2WL$

Loaded portal frame

(b)

$\lambda_p = 2 \cdot 40$

Plastic collapse state (θ small)

(c)

Plastic collapse state (θ finite)

(d)

Rotation work on rigid link

(e)

Equivalent vertical joint loading

(f)

Elastic critical mode

Figure 6.3   Behaviour of fixed-base portal frame

to RMN in Fig. 6.2, it is necessary to follow the variation in the load factor λ at increasing *finite* values of the rotation θ. At finite deformations, the mechanism itself becomes nonlinear, and a consideration of the exact geometry changes becomes laborious. It has been shown, however (see Horne[4]), that the following treatment gives a value for λ which is correct to the first power of θ.

Consider a finite state of deformation of the collapse mechanism, [Fig. 6.3(c)], and let any typical member or part of a member rotating as a rigid link (for example, AB) have undergone a rotation θ_k. Let the length of the rigid link be L_k and suppose that the axial thrust it sustains at this instant is denoted by P_k. Considering a typical hinge with plastic moment M_{pj}, let ϕ_j be the total rotation. Finally, if a typical applied load is denoted by λW_i where λ is the instantaneous load factor, let the distance through which the point of application has moved in the direction of the load be Δ_i.

Suppose now that, owing to an infinitely small additional deformation of the collapse mechanism, during which the load factor changes from λ to $(\lambda + d\lambda)$, θ_k changes to $(\theta_k + d\theta_k)$, ϕ_j to $(\phi_j + d\phi_j)$ and Δ_i to $(\Delta_i + d\Delta_i)$. The work equation for this incremental deformation may be written

$$\lambda \sum_i W_i d\Delta_i + \sum_k P_k L_k \theta_k d\theta_k = \sum_j M_{pj} d\phi_j \tag{6.5}$$

The first and last terms of the equation correspond to the external and internal work terms in simple plastic theory. The second term is due to the additional external work arising from finite deformations, and the way in which it arises may be appreciated by reference to the typical member AB in Fig. 6.3(d). The thrust P_{AB} is due primarily to the end load $P_{AB} \simeq P_{AB}$ acting in a direction parallel to the original direction of the member, and, during the additional rotation $d\theta_{AB}$, the load does an amount of work $P_{AB}(L_{AB}\theta_{AB}d\theta_{AB})$ which is not included in the first external work term in Eqn (6.5). It may be shown[4] that all additional work terms due to finite deformations are included if such terms as those just given are written for each member or part of a member rotating as a rigid link in the collapse mechanism, except for terms which are of an order of magnitude smaller than the first power of θ_k. Hence arises the second term in Eqn (6.5).

The axial thrusts P_k in Eqn (6.5) may be obtained with sufficient accuracy by proportion from the values they have in the simple collapse state. Let λ_p be the rigid–plastic collapse load factor and let the axial thrusts then be P_{kp}. It is then assumed that, at any load factor λ during the finite deformation of the collapse mechanism, the axial thrusts are $\lambda(P_{kp}/\lambda_p)$. The approximation is also made that the total and incremental rotations and displacements are all in the same proportion as those that occur for the same mechanism during an infinitely small deformation from the undeformed state. Hence

$$\frac{d\Delta_i}{\Delta_i} = \frac{d\theta_k}{\theta_k} = \frac{d\phi_j}{\phi_j}$$

and Eqn (6.5) becomes

$$\lambda \left(\sum_i W_i \Delta_i + \sum_k \frac{P_{kp}}{\lambda_p} L_k \theta_k^2 \right) = \sum M_{pj} \phi_j \tag{6.6}$$

At infinitely small deformations,

$$\lambda_p \sum_i W_i \Delta_i = \sum_j M_{pj}\phi_j \tag{6.7}$$

Eliminating $\sum_i W_i \Delta_i$ between Eqns (6.6) and (6.7),

$$\frac{\lambda}{\lambda_p} = \frac{1}{1 + \dfrac{\sum_k P_{kp} L_k \theta_k^2}{\sum_j M_{pj}\phi_j}} \tag{6.8}$$

The axial thrusts in the portal frame of Fig. 6.3(a) at collapse at the rigid–plastic collapse load factor $\lambda_p = 2\cdot40$ are readily derived (Chapter 2) and are shown in Fig. 6.3(b). Hence, for the rigid links AB, BC, CD, and DE, $P_{kp} = 5\cdot6W, 0\cdot4W, 0\cdot4W$, and $8\cdot8W$ respectively. Hence, applying Eqn (6.8),

$$\sum_k P_{kp} L_k \theta_k^2 = (5\cdot6 + 0\cdot4 + 0\cdot4 + 8\cdot8)WL\theta^2 = 15\cdot2WL\theta^2$$
$$\sum_j M_{pj}\phi_j = (\theta + 2\theta + 2\theta + \theta)M_p = 6M_p\theta = 12WL\theta$$

and $\quad \dfrac{\lambda}{\lambda_p} = \dfrac{1}{1 + \frac{19}{15}\theta} \tag{6.9}$

This gives the rigid–plastic mechanism curve RN in Fig. 6.4. The horizontal axis is proportional not only to θ, but also to the sway deflection $\Delta_H = L\theta$.

The linear elastic sway deflection is readily shown to be

$$\Delta_H = \frac{7}{32} \frac{\lambda WL^3}{EI} = \frac{7}{160} \lambda L \tag{6.10}$$

giving the straight line OKL in Fig. 6.4. This intersects the mechanism line at K at a load factor λ_K of

$$\lambda_K = 2\cdot15$$

which is therefore an upper bound on the failure load. A load factor which is slightly easier to calculate than λ_K is λ_M, obtained by finding the load on the plastic mechanism line that corresponds to the linear *elastic* deflection at the theoretical *plastic* collapse load, in this case from Eqn (6.10), when

$$\frac{\Delta_H}{L} = \frac{7 \times 2\cdot40}{160} = 0\cdot105$$

corresponding to point L in Fig. 6.4. The load factor λ_M at point M is thus found to be

$$\lambda_M = 2\cdot12$$

An improved upper bound is obtained from the nonlinear elastic curve OH. The easiest way of deriving this curve is first to obtain the *elastic critical* load factor for the portal frame. For this purpose, the vertical loads only are considered, these being assumed concentrated at the joints as shown in Fig. 6.3(e) At the first elastic critical load factor λ_E, buckling occurs as shown in Fig. 6.3(f). Methods of calculating elastic critical loads are readily available [5, 6] and make use of stability functions. If n is the no-sway stability function of Merchant for the

11—P.T.S.

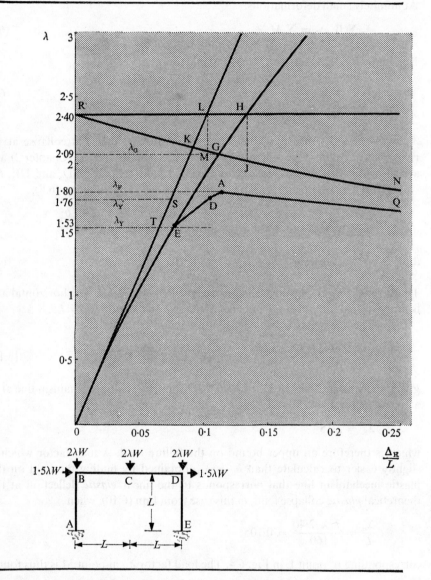

Figure 6.4   Load–deflection curves for frame in Fig. 6.3(a)

column members in Fig 6.3(e), then n is a function of the ratio ρ of the axial load in each column to the Euler critical load of each column as a pin-ended strut. Hence

$$n = \mathrm{f}(\rho) \quad \text{where } \rho = \frac{3\lambda W}{\left(\dfrac{\pi^2 EI}{L^2}\right)} = \frac{3}{5\pi^2}\lambda$$

Tables of the function n are given by Livesley and Chandler[7] and also by

Horne and Merchant.[5] Considering the bending moments M_{BA} and M_{BD} at joint B in Fig. 6.3(f), and allowing for symmetry (which makes $M_{DB} = M_{BD}$), the slope–deflection equations for members AB and BD give

$$M_{BA} = n\left(\frac{EI}{L}\right)\theta_B$$

$$M_{BD} = 6\left(\frac{EI}{2L}\right)\theta_B$$

where θ_B is the rotation at joint B. Since $M_{BA} + M_{BD} = 0$, it follows that

$$n + 3 = 0$$

The lowest value of ρ for which $n = -3$ is $\rho = 0.611$, whence the elastic critical load factor λ_E is given by

$$\lambda_E = \frac{5\pi^2}{3}(0.611) = 10.04$$

The elastic critical load is necessarily an upper bound on the failure load, and even for such a slender structure as that being considered, this upper bound lies well above the upper bound given by the rigid–plastic load ($\lambda_p = 2.40$).

A close approximation to the nonlinear elastic load–deflection curve may now be obtained by applying to the linear deflections the amplification factor $1/\{1 - (\lambda/\lambda_E)\}$ (see, for example, Horne and Merchant[5]). This leads, using Eqn (6.10), to

$$\Delta_H = \frac{7}{160}\left(\frac{\lambda\lambda_E}{\lambda_E - \lambda}\right)L \tag{6.11}$$

which is represented graphically by OEH in Fig. 6.4. The curve rises to meet the elastic critical load line $\lambda = \lambda_E = 10.04$ asymptotically. The point of intersection G with the mechanism line gives an upper bound to the failure load

$$\lambda_G = 2.09$$

while the point J, at a deflection equal to the nonlinear elastic deflection for $\lambda = \lambda_p = 2.40$, gives

$$\lambda_J = 2.04$$

This load factor, it may be noted, is not necessarily an upper bound and would not be so, for example, for the eccentrically loaded column for which results are given in Fig. 6.2.

The load at which yield first occurs on the basis of linear elastic analysis is readily obtained, and is found to be at base E (Fig. 6.3) where the moment is given by $M_E = \frac{91}{80}\lambda WL$. The load factor λ'_Y at which $M_E = M_p = 2WL$ is therefore

$$\lambda'_Y = \frac{160}{91} = 1.76$$

represented by point S in Fig. 6.4. If the shape factor for the cross section is 1·15, the load factor λ_Y for first yield becomes

$$\lambda_Y = \frac{1·76}{1·15} = 1·53$$

These calculations represent the limit of simple hand methods, and establish that the true load-deflection curve lies below OGN in Fig. 6.4. It is possible, assuming unit shape factor and making use of stability functions, to follow step by step the formation of plastic hinges and thereby to derive the theoretical elastic–plastic load–deflection curve. The result for the present frame is OEDAQ in Fig. 6.4. Computer programs have been produced to carry out such analyses,[8–10] although a lot of computer time is required for large frames because of the necessity to recalculate and invert the stiffness matrix for the structure after the formation of each plastic hinge, or at least after each few additional hinges have formed. The collapse load for the present structure is found to be $\lambda_F = 1·80$, representing a 25% reduction below the simple plastic collapse load $\lambda_p = 2·40$. Allowance for the spread of plasticity away from plastic hinge positions would cause the theoretical curve to fall very slightly below the curve EDAQ, but the effect of this on the theoretical failure load is negligible.

In the unit shape factor analysis, point E in Fig. 6.4 represents the load factor ($\lambda = 1·55$) at which the first plastic hinge forms, this being at base E. The second hinge forms at D, the third at A, and the fourth at B, at a load factor $\lambda = 1·57$ and deflection $\Delta_H = 0·36L$ (beyond point Q in Fig. 6.4). The final

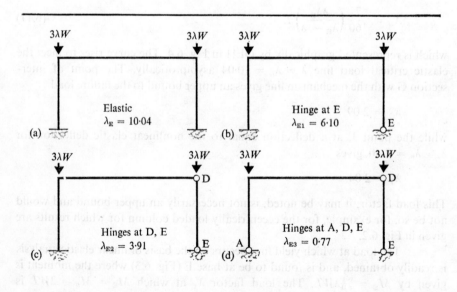

Figure 6.5 Deteriorated critical loads for frame in Fig. 6.3(a)

mechanism, therefore, differs from the combined mechanism, by which the frame fails according to simple plastic theory. This, however, is of no importance since the peak load occurs at the formation of the third hinge at A, and failure actually takes place due to the deterioration in the stiffness of the structure without the formation of a complete mechanism.

The progressive deterioration in the stiffness of a structure with the successive formation of plastic hinges has been discussed by Wood.[11] He pointed out that, after the first hinge has formed, there is a reduction in the critical load to a deteriorated value. Thus, after the formation of the plastic hinge at E, the load–deflection curve ED in Fig. 6.4 is part of a curve which, if no further hinges formed, would become asymptotic to the line $\lambda = \lambda_{E1} < \lambda_E$ where λ_{E1} is the elastic critical load of the deteriorated structure in Fig. 6.5(b). This has a structural hinge at E, representing the elimination of all stiffness at this point due to the plastic hinge. With the second hinge formed at D, the deteriorated structure [Fig. 6.5(c)] has hinges at D and E, and has a critical load $\lambda_{E2} < \lambda_{E1}$. With the third deteriorated structure, which has hinges at D, E, and A [Fig. 6.5(d)], the elastic critical load becomes *less* than the load factor at A in Fig. 6.4, thus causing collapse. The physical picture is therefore that of the progressive formation of hinges accompanied by successive decreases in the deteriorated critical load, until that stage is reached at which the deteriorated critical load falls below the current load level. Collapse then takes place.

6.4   Practical estimation of failure loads

The calculations in Section 6.3 are too elaborate to be carried out for the average structure, particularly in view of the fact that actual reductions in failure loads below the simple plastic values are likely to be very much less than the 25% found for the portal frame in Fig. 6.3. As an upper bound, the load corresponding to any one of the points K, M, G, or J in Fig. 6.4 may be used, although points M and J are not upper bounds in all cases. Point M is probably the most convenient, since it involves finding merely the linear elastic deflection at the simple plastic collapse load, and then the corresponding load factor on the rigid–plastic mechanism curve [Eqn (6.8)].

The point S on the linear load–deflection curve at which a plastic hinge first forms is not strictly a lower bound, but is found to be so for all practical structures. In many structures it is, however, considerably below the failure load, and may be as much as 40% or more below.

Since upper and lower bounds derived as just described will not be sufficiently close together for the majority of structures, semi-empirical procedures are necessary to obtain realistic estimates of the failure load whenever it is anticipated that this may fall appreciably below the simple plastic value. An empirical formula which has, however, been shown to have some theoretical justification[4] is the Rankine–Merchant[5,12] load factor λ_R, where

$$\frac{1}{\lambda_R} = \frac{1}{\lambda_p} + \frac{1}{\lambda_E}$$

or $\qquad \lambda_R = \dfrac{\lambda_p}{1 + \dfrac{\lambda_p}{\lambda_E}}$ $\hfill$ (6.12)

This failure load is therefore estimated from the two parameters λ_p, the rigid–plastic load factor which effectively ignores elasticity, and λ_E, the elastic critical load factor which ignores plastic deformation. Figure 6.6 shows experimental failure load factors λ_F for a number of model 3, 5, and 7-storey frames tested by Low.[13] The values of λ_F are plotted in a space defined by axes λ_F/λ_E and λ_F/λ_p, where λ_E and λ_p are the theoretical elastic critical and rigid plastic loads respectively. The Rankine–Merchant formula 6.12 is represented by the straight line AB, and it is seen that the formula does represent for these frames a lower bound on the failure load. Theoretical values of the failure load factor λ_F were calculated for a large number of single and two-storey single bay frames by Salem[14,15] and, while λ_F is slightly below λ_R for some of these frames, the Rankine–Merchant load is shown to represent an approximate lower bound. The fact that, in Low's tests, no frames failed below the Rankine–Merchant load may perhaps be attributed to the effect of strain-hardening.[16] Moreover, λ_R appears to be a safe estimate of the failure load for frames with small or moderate side loads. In Low's tests, frames with very small side loads failed in some cases well above λ_R. The Rankine–Merchant load only becomes unsafe for very high side loads—normally higher than those involved for building frames. This is illustrated by the frame in Fig. 6.3, for which

$$\lambda_R = \frac{2\cdot40}{1 + \dfrac{2\cdot40}{10\cdot04}} = 1\cdot93$$

compared with an actual failure load of $1\cdot80$. Korn[17] has also investigated the application of the Rankine–Merchant load to a number of multi-storey frames, and shows that an approximate estimate of the elastic critical load is sufficient for use in Eqn (6.12).

Another semi-empirical procedure for estimating the effect of deflections on the failure loads of rigid frames is to take a point on the rigid–plastic mechanism line corresponding to a deflection somewhat in excess of that calculated from linear elastic behaviour at the plastic collapse load. Thus, in the case of the portal frame in Fig. 6.3, the failure load factor of $1\cdot80$ corresponds to a deflection on the rigid–plastic line RN in Fig. 6.4 approximately $2\frac{1}{2}$ times that of point L. Examination of a number of cases shows that this procedure usually gives a safe but fairly close estimate of the actual value.

Experiments show that the majority of frames actually collapse at load factors slightly in excess of theoretical plastic collapse loads, despite the effect of change of geometry. This is because of the influence of strain-hardening[16,18] which is usually found to compensate for theoretical reduction of carrying capacity due to instability, provided the Rankine–Merchant load [Eqn

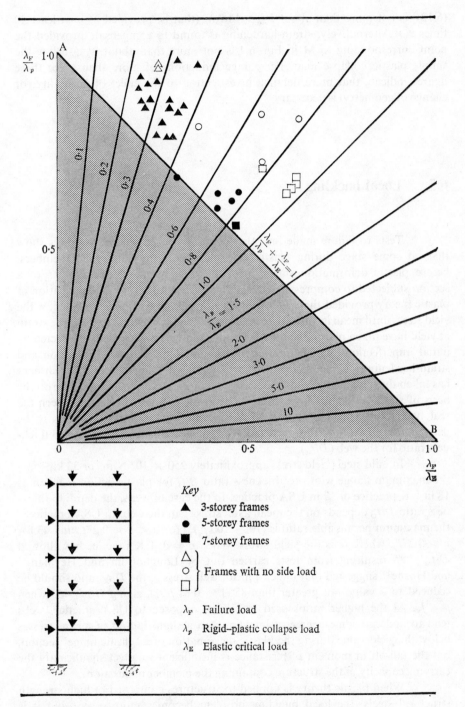

Figure 6.6 Failure loads of multi-storey frames tested by Low

(6.12)] is not more than about 5% below λ_p (that is, provided λ_E is at least 20 times λ_p). Alternatively, strain-hardening is found to compensate provided the point corresponding to M in Fig. 6.4 is not more than about $2\frac{1}{2}$% below the simple plastic collapse load. Percentage reductions of more than twice these figures indicate that more detailed investigation of the effect of instability (or change of geometry) is necessary.

6.5   Local buckling

Tests to failure made on either open- or closed-section members show that, at some stage during plastic deformation or, in thin-walled members, before plastic deformation starts, local buckling occurs in those parts of a section subjected to compressive stress. This does not prevent the application of plastic theory provided there is no fall-off in mean stress significantly below the yield stress until mean in-plane deformations equal to two or three times the strain at yield have taken place. The post-buckling load-carrying capacity is affected by initial imperfections, initial internal stresses and the plastic deformation and strain-hardening characteristics of the material, and no theoretical treatment has taken these all into account. Extensive theoretical work has been done on the basis of strain-hardening behaviour,[19] but in the end test results have been the real basis of practical proposals. The most important thickness ratios to be defined are the width to thickness ratio for flanges (B/t_f) and the depth to thickness ratio for the web (D/t_w).

In mild steel (yield stress approximately 250×10^6 N/m^2 or 33 kips/in^2) the maximum flange width to thickness ratio B/t_f for plastic action is taken as 18 in UK practice or 17 in USA practice. In the case of webs, the depth to thickness ratio D/t_w depends on the mean axial stress p in the web. In USA practice[20] the maximum permissible ratio is $(70 - 100\ p/f_y)$ for $0 < p < 0.27f_y$ and 43 for $p > 0.27f_y$, where f_y is the yield stress. When $p = 0$, UK practice has allowed $D/t_w = 85$ resulting from tests carried out by Longbottom and Heyman,[21] and Horne[22] suggested that under a mean web stress p, the D/t_f ratio should be reduced to a value not greater than $85/(1 + 0.61\ p/f_y)$, giving $D/t_f = 53$ when $p = f_y$. At the higher ratios used in UK as opposed to USA practice, webs tend to undergo some degree of buckling and a slight fall-off in mean stresses below the yield value during the plastic deformation of a 'plastic hinge' section, but the fall-off in moment of resistance is insufficient to affect significantly the carrying capacity of the structure containing the member in question.

When plastic theory is applied to structures composed of high-strength structural steels, the local buckling problem becomes more severe and it is necessary to multiply the above limiting ratios by $\sqrt{\{(250 \times 10^6)/f_y\}}$ (with f_y in N/m^2) or $\sqrt{(33/f_y)}$ (with f_y in kips/in^2).

Bibliography

1   Horne, M. R. The elastic-plastic theory of compression members. *J. Mech. Phys. Solids*, 1956, **4**, 104.

2   Murray, N. W. The determination of the collapse loads of rigidly jointed frameworks with members in which axial forces are large. *Proc. Instn civ. Engrs*, 1956, **5** [III], 213.

3   Murray, N. W. Further tests on braced frameworks. *Proc. Instn civ. Engrs*, 1958, **10**, 503.

4   Horne, M. R. Elastic-plastic failure loads of plane frames. *Proc. R. Soc.*, 1963, **274**[A], 343.

5   Horne, M. R. and W. Merchant. *The Stability of Frames*, Pergamon Press, Oxford, 1965.

6   Gregory, M. *Elastic Instability*, E. and F. Spon, London, 1967.

7   Livesley, R. K. and D. B. Chandler. *Stability Functions for Structural Frameworks*, Manchester University Press, Manchester, 1956.

8   Livesley, R. K. The application of computers to problems involving plasticity. Symposium on the use of electronic computers in structural engineering, University of Southampton, 1959.

9   Jennings, A. and K. I. Majid. An elastic plastic analysis by computer for framed structures loaded up to collapse. *Structural Engineer*, 1965, **43**, 407.

10   Majid, K. I. and D. Anderson. The computer analysis of large multi-storey framed structures. *Structural Engineer*, 1968, **46**, 357.

11   Wood, R. H. Stability of tall buildings, *Proc. Instn civ. Engrs*, 1968, **11**, 69.

12   Merchant, W. The failure load of rigidly jointed frameworks as influenced by stability. *Structural Engineer*, 1954, **32**, 185.

13   Low, M. W. Some model tests on multi-storey rigid steel frames. *Proc. Instn civ. Engrs*, 1959, **13**, 287.

14   Salem, A. PhD Thesis, Manchester University, 1958.

15   Horne, M. R. The stability of elastic-plastic structures. *Progress in Solid Mechanics*, North Holland Publishing Co., Amsterdam, 1961.

16   Horne, M. R. Instability and the plastic theory of structures. *Trans. Eng. Inst., Canada*, 1960, **4**, 31.

17   Korn, A. *The approximation of stability effects on frames*. International Association of Bridge and Structural Engineering, 1968, 101.

18   Horne, M. R. Plastic design of portal frames in steel to BS 968. British Constructional Steelwork Association, *Publication No 29*, 1966.

19   Haaijer, G. Plate buckling in the strain-hardening range. *Trans. Am. Soc. civ. Engrs*, 1959, **124**, 117.

20   *Commentary on Plastic Design*. ASCE Manual No. 41, WRC–ASCE Joint Committee, 1961.

21   Longbottom, E. and J. Heyman. Experimental verification of the strengths of plate girders designed in accordance with the revised

British Standard 153: tests on full-size and on model plate girders. *Proc. Instn civ. Engrs*, 1956, **5**[III], 462.

22   Horne, M. R. The full plastic moments of sections subjected to shear force and axial loads. *British Welding Journal*, 1958, **5**, 170.

Problems

6.1   A beam-column AB of length L is completely fixed in direction at A and B and carries a uniformly distributed lateral load W together with an axial thrust $2W$. The section of the beam has a depth $d = L/30$, an area A, a second moment of area about the axis of bending of $Ad^2/6$ and a plastic modulus under zero axial thrust of $\frac{3}{8}Ad$. The yield stress is f_y and the elastic modulus $E = 900f_y$. When the ratio of mean axial stress to yield stress is n, the plastic modulus becomes $\frac{3}{8}Ad(1 - 1\cdot5n^2)$ provided $0 \leqslant n \leqslant 0\cdot40$ and $\frac{1}{24}Ad\,(1 - n)\,(11 + n)$ when $0\cdot40 \leqslant n \leqslant 1\cdot00$.

(a)   Determine the value of n at rigid–plastic collapse.

(b)   Obtain the formula for the mechanism line RN in Fig. 6.4, plotting n against Δ/d where Δ is the central deflection.

(c)   Show that the elastic critical load level under axial thrust $2W$ occurs when $W = \pi^2 Af_y/3$.

(d)   Obtain an approximation to the nonlinear elastic curve OEH in Fig. 6.4 (n versus Δ/d), given that the linear elastic deflection is $WL^3/384EI$.

(e)   Obtain an estimate of the collapse load as given by point G in Fig. 6.4, expressing the result as a percentage reduction below the rigid–plastic load.

(f)   Obtain similarly the Rankine–Merchant load.

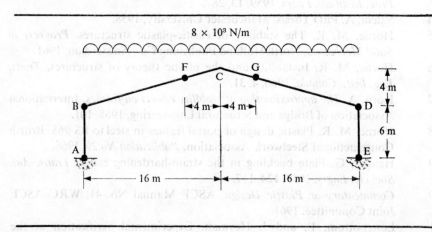

Figure E6.1

6.2   Obtain the variation of load W with the deflection at the top of the structure during the rigid–plastic collapse of the frame in Example 2.6 (see answers for simple rigid–plastic collapse mode). Ignore the effect of axial thrusts in the beams.

6.3   Obtain the variation of load factor λ with the deflection at the top of the structure during the rigid–plastic collapse of the frame in Example 2.7. Ignore the effect of axial thrusts in the beams.

6.4   The fixed-base, pitched-roof portal frame in Fig. E6.1 has a uniform section with a plastic moment of 5×10^5 Nm and is acted upon by a uniformly distributed vertical load of 8×10^5 N/m, measured on plan. If the simple plastic collapse mechanism has plastic hinges in the positions shown in Fig. E6.1, derive the rigid–plastic relationship between the load factor λ and the eaves deflection Δ, where Δ is expressed in metres. Assume that the frame displacements are symmetrical, and that the effect of thrust on the plastic moment may be neglected.

The eaves deflection at unit load factor, calculated by linear elastic theory, is 60 mm. Use this fact to derive an estimate of the failure load, and comment on the likely accuracy of your result.

Answers to problems

Chapter 1

1.1   (a) $0.061d^3$, 1.240
         (b) $0.0145d^3$, 1.688
         (c) $d^3/6$, $16/3\pi = 1.697$
         (d) td^2, $4/\pi = 1.274$
         (e) $d^3/3\sqrt{2}$, 2
         (f) $1.5td^2$, 1.125
         (g) $\sqrt{2}\, td^2$, 1.5

1.2   18 kNm. Collapse in span CD

1.3   At mid-span. $24M_p/L^2$

1.4   (a) $w = 8(3 + 2\sqrt{2})M_p/L^2 = 46.6M_p/L^2$
         (b) $w = 8(1 + n\sqrt{2})^2 M_p/L^2$
             End spans $(1 + \sqrt{2})L/2(1 + n\sqrt{2})$
             Intermediate spans $\sqrt{2}\{L/(1 + n\sqrt{2})\}$

1.5 $L/(5 + \sqrt{17}) = 0.1096L$, $(5 + \sqrt{17})\,M_p/L = 9.12M_p/L$

1.6 $1.5M_p/L$

1.9

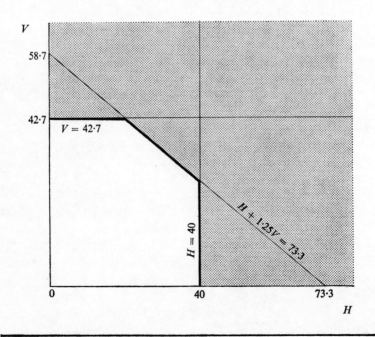

1.10

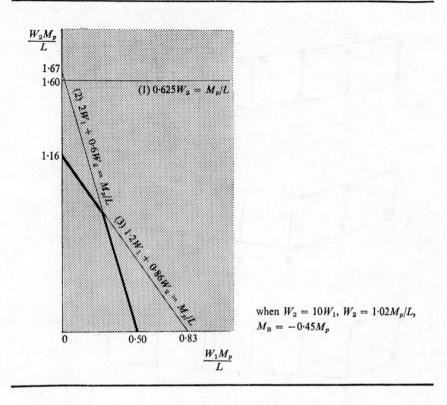

(1) $0.625W_2 = M_p/L$

(2) $2W_1 + 0.6W_2 = M_p/L$

(3) $1.2W_1 + 0.86W_2 = M_p/L$

when $W_2 = 10W_1$, $W_2 = 1.02M_p/L$,
$M_B = -0.45M_p$

1.11

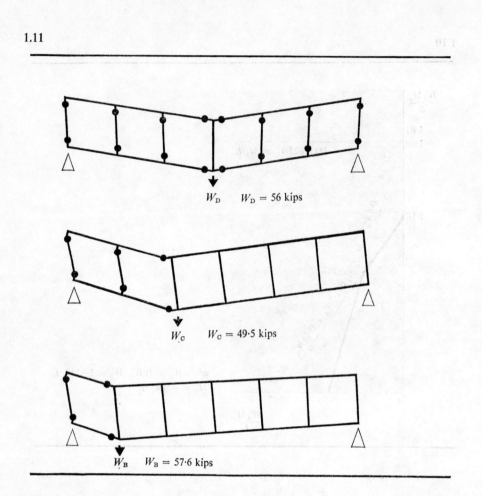

W_{D} $W_{\mathrm{D}} = 56$ kips

W_{C} $W_{\mathrm{C}} = 49 \cdot 5$ kips

W_{B} $W_{\mathrm{B}} = 57 \cdot 6$ kips

1.12

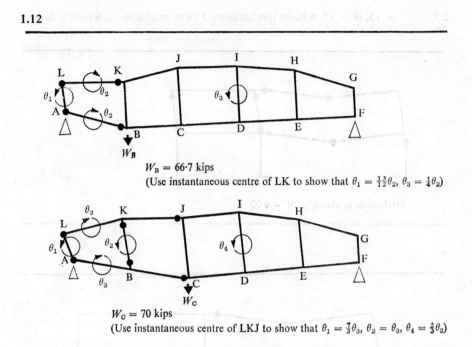

$W_B = 66.7$ kips

(Use instantaneous centre of LK to show that $\theta_1 = \frac{13}{12}\theta_2$, $\theta_3 = \frac{1}{4}\theta_2$)

$W_C = 70$ kips

(Use instantaneous centre of LKJ to show that $\theta_1 = \frac{7}{3}\theta_3$, $\theta_2 = \theta_3$, $\theta_4 = \frac{2}{3}\theta_3$)

Chapter 2

2.2   30 kNm, 22·5 kNm, 37·5 kNm
2.3   40 kNm, 32·7 kNm, 43·6 kNm
2.4

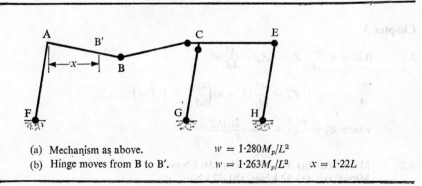

(a) Mechanism as above. $w = 1\cdot280 M_p/L^2$
(b) Hinge moves from B to B'. $w = 1\cdot263 M_p/L^2$ $x = 1\cdot22L$

2.5 $r = 15, m = 15$, 6 beam mechanisms, 3 sway mechanisms, 6 joint rotations.
2.6

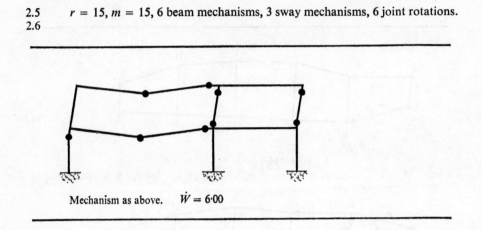

Mechanism as above. $\dot{W} = 6{\cdot}00$

2.7

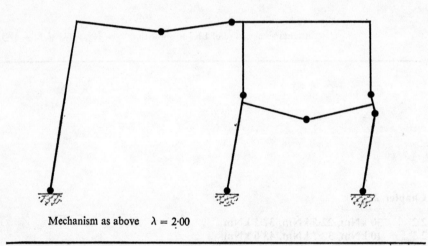

Mechanism as above $\lambda = 2{\cdot}00$

Chapter 3

3.1 $0 \leqslant n \leqslant \dfrac{dt_w}{A},\; Z'_p = Z_p - \dfrac{A^2}{4d}\, n^2$

$\dfrac{dt_w}{A} \leqslant n \leqslant 1,\; Z'_p = \dfrac{A^2}{8t_f}(1 - n)\!\left(\dfrac{4bt_f}{A} - 1 + n\right)$

where $Z_p = \dfrac{bt_f{}^2}{2} + \dfrac{(d - 2t_f)t_w{}^2}{4}$

3.2   Major axis (a) 370 kNm, (b) 156 kNm
        Minor axis (a) 62 kNm, (b) 47 kNm

3.3 $Z_p' = \dfrac{bd^2}{3\sqrt{2}}\,(1 - n)\,\{\sqrt{2} - \sqrt{(1 - n)}\}$,

$n = \tfrac{1}{9},\ Z_p' = \tfrac{8}{81}bd^2$

3.4 $W = \dfrac{2bL}{H}\,\{\sqrt{(L^2 + d^2)} - L\}\,\sigma_y$

3.5   Sagging moment:

$-1 \leqslant n \leqslant \tfrac{1}{3},\ M = (1 + n)(7 - 9n)td^2\sigma_y/4$

$\tfrac{1}{3} \leqslant n \leqslant 1,\ M = 2(1 - n)td^2\sigma_y$

Hogging moment:

$-1 \leqslant n \leqslant -\tfrac{1}{3},\ M = -2(1 + n)td^2\sigma_y$

$-\tfrac{1}{3} \leqslant n \leqslant 1,\ M = -(1 - n)(7 + 9n)td^2\sigma_y/4$

$M_{\max} = 16\,td^2\sigma_y/9$ when $n = -\tfrac{1}{9}$

3.6   474 kN, 1·9%

Chapter 4

4.1 $M_1 = \tfrac{7}{6}\,WL,\ M_2 = \tfrac{2}{3}WL$

4.2   For $\dfrac{W_1L_1}{W_2W_2} \leqslant 0\cdot75$

$\dfrac{L_1}{L_2} \leqslant 0\cdot5,\ M_1 = M_2 = \dfrac{W_2L_2}{8}$

$\dfrac{L_1}{L_2} \geqslant 0\cdot5,\ M_1 = \dfrac{W_1L_1}{6},\ M_2 = \dfrac{W_2L_2}{4} - \dfrac{W_1L_1}{6}$

For $\dfrac{W_1L_1}{W_2L_2} \geqslant 0\cdot75$

$\dfrac{L_1}{L_2} \leqslant 1,\ M_1 = \dfrac{W_1L_1}{4} - \dfrac{W_2L_2}{16},\ M_2 = \dfrac{W_2L_2}{8}$

$\dfrac{L_1}{L_2} \geqslant 1,\ M_1 = M_2 = \dfrac{W_1L_1}{6}$

4.3 $M_1 = 4\cdot5L,\ M_2 = 7\cdot5L,\ M_3 = 3L,\ M_4 = 4\cdot5L$

4.4   Permissible region bounded by
       (1) $M_1 = 3\cdot75WL$ (loading A)
       (2) $2M_1 + M_2 = 15\cdot5WL$ (loading A)
       (3) $M_1 + M_2 = 11WL$ (loading B)
       (4) $M_2 = 5\cdot5WL$ (loading B)
       Minimum weight is at intersection of (2) and (3), that is,
       $M_1 = 4\cdot5WL,\ M_2 = 6\cdot5WL$

Chapter 5

5.1 $W = \dfrac{4L^2(L+a)}{a(L-a)(4L^2+La+a^2)} M_p$

5.2 $W = 3 \cdot 86 M_p/L$

5.3   (a) $6 \cdot 4 M_p/L$, (b) $4 \cdot 99 M_p/L$

5.4 $1 \cdot 610$, hinges at sections 1, 3, 4, 6, 8, 9, 10. Residual moments at sections 1 to 10 are respectively, -86, -92, -9, -74, 70, -16, 4, -4, 13, -47.

5.5   (a) $1 \cdot 493$, (b) $1 \cdot 347$

Chapter 6

6.1   (a) $\frac{1}{3}$

         (b) $\dfrac{\Delta}{d} = \dfrac{0 \cdot 75}{n} - 1 \cdot 125n - 1 \cdot 875$

         (d) $\dfrac{\Delta}{d} = 15n/64 \left(1 - \dfrac{3}{2\pi^2} n\right)$

         (e) $3 \cdot 0\%$

         (f) $5 \cdot 0\%$

6.2 $W = 6/(1 + \frac{3}{155}\Delta)$

6.3 $\lambda = 2/\left(1 + \dfrac{\Delta}{16}\right)$

6.4 $\lambda = 1 \cdot 5625/(1 + 0 \cdot 595\Delta)$
         $\lambda = 1 \cdot 480$

Index